Machinery Failure Analysis Handbook

Machinery Failure Analysis Handbook

Editor

Promita Mukharjee

scitus
academics

Machinery Failure Analysis Handbook
Edited by **Promita Mukharjee**

Printed in 2017

ISBN: 978-1-68117-409-9

Library of Congress Control Number: 2015941616

© 2016 by
SCITUS Academics LLC,
616, Corporate Way, Suite 2, 4766,
Valley Cottage, NY 10989

www.scitusacademics.com

Notice

Contents

Preface

Understanding why and how failures occur is critical to failure prevention, as even the slightest breakdown can lead to catastrophic loss of life and asset as well as widespread pollution. This book helps anyone involved with machinery reliability, whether in the design of new plants or the maintenance and operation of existing ones, to understand why process equipment fails and thereby prevent similar failures. Process industries are home to a huge number of machines, most of them critical to the industry mission. Failures of these machines can result in consequences that range from the simple replacement of a cheap bearing to an accident that may cost millions in lost production or cause injuries or pollution. Competition forces corporations to try to keep a pace in optimization. On the machinery side of the history, that means improving efficiencies, reliability, and reducing maintenance cost. Designs and purchase specifications, shop testing, installation, maintenance and operation all play a role in these efforts. The objective of this book is to help anyone involved with machinery reliability, be it in the design of new plants or the maintenance and operation of existing ones, to understand why the process machine fails, so some preventive measures can be taken to avoid another failure of the same kind.

Editor

Machine Fault Signature Analysis

Pratesh Jayaswal[1], A. K. Wadhwani[2], and
K. B. Mulchandani[3]

[1]Department of Mechanical Engineering, Madhav Institute of Technology and Science, Gwalior 474005, India

[2]Department of Electrical Engineering, Madhav Institute of Technology and Science, Gwalior 474005, India

[3]Department of Mechanical and Industrial Engineering, Indian Institute of Technology Roorkee, Roorkee 247667, India

ABSTRACT

The objective of this paper is to present recent developments in the field of machine fault signature analysis with particular regard to vibration analysis. The different types of faults that can be identified from the vibration signature analysis are, for example, gear fault, rolling contact bearing fault, journal bearing fault, flexible coupling faults, and electrical machine fault. It is not the intention of the authors to attempt to provide a detailed coverage of all the faults while detailed consideration is given to the subject of the rolling element bearing fault signature analysis.

INTRODUCTION

Machine fault problems are broad sources of high maintenance cost and unwanted downtime across the industries. The prime objective of maintenance department is to keep machinery and plant equipments in good operating condition that prevents failure and production loss. If the department organizes a predictive maintenance program, this goal as well as cost benefits can be achieved, while accurate information at the right time is a crucial aspect of a maintenance regimen [1]. The condition-based maintenance strategy is being employed for uninterrupted production process in industries. Condition-based maintenance (CBM) consists of continuously evaluating the condition of a monitored machine and thereby successfully identifying faults before catastrophic breakdown occurs. Numerous condition monitoring (cm) and diagnostics methodologies are utilizing to identify the machine faults to take corrective action. Machine fault identification can be done with different methodologies as vibration signature analysis, lubricant signature analysis, noise signature analysis, and temperature monitoring, with the use of appropriate sensors, different signal conditioning, and analyzing instruments.

Vibration signature analysis techniques for machine fault identification are the most popular among other techniques. Vibration monitoring is based on the principle that all the system produces vibration. When a machine is operating properly, the vibration is small and constant, however, when faults develop and some of the dynamic process in the machine changes, there will be changes in vibration spectrum observed. After the review of previous published work, it is concluded that gear fault, bearing fault, and coupling fault are studied for research purpose to fault signature analysis. The majority of industrial machines use ball or rolling elements bearings (REB). The vibration signals obtained from the vicinity of a bearing assembly contain rich information about the bearing condition. Most of the researchers have used vibration signature analysis techniques for rolling element bearing fault identification in case of single defect on bearing components. Time-domain and frequency-domain vibration analysis techniques were tested but effective identification of bearing condition is, however, not so straightforward. Several researchers have used artificial intelligence techniques as well as time-frequency domain analysis and developed expert diagnostics system for bearing fault identification with the use of artificial neural network, fuzzy logic, wavelet transform, and hybrid techniques. In this paper, a review of need and different techniques of machine fault signature analysis are discussed. A special emphasis is given to rolling element bearing vibration signature analysis, while other techniques are also covered. This paper is divided into different sections, each dealing with various aspects of the subject. It begins with a summary of need of machine fault diagnosis followed by a general overview of the numerous means of signature analysis.

NECESSITY OF MACHINE FAULT IDENTIFICATION

Machine fault can be defined as any change in a machinery part or component which makes it unable to perform its function satisfactorily or it can be defined as the termination of availability of an item to perform its intended function. The familiar stages before the final fault are incipient fault, distress, deterioration, and damage, all of them eventually make the part or component unreliable or unsafe for continued use [2]. Classification of failure causes are as follows:

- inherent weakness in material, design, and manufacturing
- misuse or applying stress in undesired direction

- gradual deterioration due to wear, tear, stress fatigue, corrosion, and so forth.Antifriction bearings failure is a major factor in failure of rotating machinery. Antifriction bearing defects may be categorized as localized and distributed. The localized defects include cracks, pits, and spalls caused by fatigue on rolling surfaces. The distributed defect includes surface roughness, waviness, misaligned races, and off-size rolling elements. These defects may result from manufacturing and abrasive wear [3].

Modern manufacturing plants are highly complex. Failure of process equipments and instrumentation increased the operating costs and resulted in loss of production. Undetected or uncorrected malfunctions can induce failures in related equipments and, in extreme cases, can lead to catastrophic accidents. Early fault detection in machines can save millions of dollars on emergency maintenance and production-loss cost. Gearbox and bearings are essential parts of many machineries [4]. The early detection of the defects, therefore, is crucial for the prevention of damage and secondary damage to other parts of a machine or even a total failure of the associated large system can be triggered [5].

There are certain objectives of machine fault identification:

- prevention of future failure events
- assurance of safety, reliability, and maintainability of machineries.

Machineries failures reveal a reaction chain of cause and defect. The end of the chain is usually a performance deficiency commonly referred to as the symptom, trouble, or simply the problem. The machine fault signature analysis works backwards to define the elements of the reaction chain and then proceeds to link the most probable failure cause based on failure analysis with a root cause of an existing or potential problem. Accurate and complete knowledge of the causes responsible for the breakdown of a machine is necessary to the engineer, similarly, as knowledge of a breakdown in health is to the physician. The physician cannot assure a lasting cure unless he knows what lies at the root of the trouble, and the future usefulness of a machine often depends on correct understanding of the causes of failure. The proper maintenance can be done only after the knowledge of root cause of failure.

Edwards et al. [6] present a review on fault diagnosis of rotating machinery to provide a broad review of the state of the art in fault diagnosis techniques. The early fault detection and diagnosis allow preventive maintenance and condition-based maintenance to be arranged for the machine during scheduled period of downtime caused by extensive system failures that improves the overall availability, performance and reduces maintenance cost. For the fault diagnosis problem, it is not only to detect fault in system, but also to isolate the fault and find out its causes.

CONDITION-BASED MAINTENANCE

Maintenance is a combination of science, art, and philosophy. The rationalization of maintenance requires a deep insight into what maintenance really is. Efficient maintenance is a matter of having the right resources in the right place at the right time. Maintenance can be defined as the total activities carried out in order to restore or renew an item to working condition, if fault is there. Maintenance is also defined as combination of action carried out to return an item to or restore it to an acceptable condition. The classification of maintenance according to timings of action for maintenance is shown in Table 1.

Table 1: Timings of action for maintenance

Timings of action	Maintenance
Operating to failure	Shutdown or breakdown
Fixed time based	Preventive
Condition based	Predictive or diagnostic

Every machine component behaves as an individual. Failure can take place earlier or later than recommended in case of preventive maintenance. It can be improved by condition-based maintenance. Dileo et al. [7] present a review on the classical approaches to maintenance and then compare them with condition-based maintenance (CBM).

The prevention of potential damage to machinery is necessary for safe, reliable operation of process plants. Failure prevention can be achieved by sound specification, selection, review, and design audit routines. When failures do occur, accurate definition of root cause is an absolute prerequisite to the prevention of future failure events [2].

Condition-based maintenance is defined as "maintenance work initiated as a result of knowledge of the condition of an item from routine or continuous checking." It is carried out in response to a significant deterioration in a unit as indicated by a change in a monitored parameter of the unit condition or performance. Condition reports arise from human observations, checks, and tests, or from fixed instrumentation or alarm systems grouped under the name condition monitoring. It is here that one can make use of predictive maintenance by using a technique called signature analysis. Signature analysis technique is intended to continually monitor the health of the equipment by recording systematic signals or information derived from the form of mechanical vibrations, noise signals, acoustic and thermal emissions, change in chemical compositions, smell, pressure, relative displacement, and so on [8]. Mann et al. [9] present an article explores the benefits of condition-based preventive maintenance compared to the traditional statistical reliability approach. Nandi and Toliyat [10] present a review on condition monitoring and fault diagnosis of electrical machines. Marcus [11] proposed condition-based maintenance to rail vehicle for more effective maintenance.

Condition-based maintenance differs from both failure maintenance and fixed-time replacement. It requires monitoring of some condition-indicating parameter of the unit being maintained. This contrasts with failure maintenance, which implies that no successful condition monitoring is undertaken and with fixed-time replacement which is based on statistical failure data for a type of unit. In general, condition-based maintenance is more efficient and adaptable than either of the other maintenance actions. On indication of deterioration, that unit can be scheduled for shutdown at a time chosen in advance of failure, yet, if the production policy dictates, the unit can be run to failure. Alternatively, the amount of unnecessary preventive replacement can be reduced, while if the consequences of failure are sufficiently dire, the condition monitoring can be employed to indicate possible impending failure well before it becomes a significant probability. The trend monitoring method for one or group of similar machines is possible if sufficient data of monitored parameters are available. It relates the condition of machine(s) directly to the monitored parameters. On the other hand, condition checking method is employed for a wide range of diagnostics instruments apart from human senses. Some of the recent developments in the form of CBM are proactive maintenance, reliability centered maintenance (RCM) and total productive maintenance (TPM).

MACHINE CONDITION MONITORING

When a fault takes places, some of the machine parameters are subjected to change. The change in the machine parameters depends upon the degree of faults and the interaction with other parameters. In most cases, more than one parameter are subjected to change under abnormal condition. Condition monitoring can be carried out when the equipment is in operation, which known as on-line, or when it is off-line, which means when it is down and not in the operation. While on-line, the critical parameters that are possible to monitor are speed, temperature, vibration, and sound. These may be continuously monitored or may be done periodically. Off-line monitoring is carried out when the machine is down for whatever reason. The monitoring in such would include crack detection, a thoroughly check of alignment, state of balancing, the search for tell-tale sign of corrosion, pitting, and so on.

The International Standards Organization's Technical Committee 108 (ISO/TC108) produces standards in the area of mechanical vibration, shock, and machine condition monitoring. ISO/TC108's Subcommittee 5 (ISO/TC108/SC5) has focused on standards for the condition monitoring and diagnostic of machines. This subcommittee has published ISO 13374-1:2003 which establishes general guidelines for software specifications related to data processing, communication, and presentation of machine condition monitoring and diagnostic information. This standard defines the data processing and information flow needed between processing blocks in condition monitoring systems. Machine condition monitoring (MCM) is a vital component of preventive and predictive maintenance programs that seek to reduce cost and avoid unplanned downtime. MCM also contributes to health and safety by recognizing faults which may give rise to pollution or health hazards, and also by indication of incipient faults which could produce danger conditions. MCM setups include measurement hardware and software that acquire and interpret signals generated by the machine being monitored. Condition monitoring is taken to mean the use of advanced technologies in order to determine equipment condition, and to predict potential failure. It includes, but is not limited to, technologies such as visual inspection, vibration measurement and analysis, temperature monitoring, acoustic emission analysis, noise analysis, oil analysis, wear debris analysis, motor current signature analysis, and nondestructive testing.

Visual Inspection

Visual monitoring can sometimes provide a direct indication of the machine's condition without the need for further analysis. The available techniques can range from using a simple magnifying glass or low-power microscope. Other forms of visual monitoring include the use of dye penetrants to provide a clear definition of any cracks occurring on the machine surface, and the use of heat-sensitive or thermographic paints. The condition of many transmission components can readily be checked visually. For example, the wear on the surfaces of gear teeth gives much information. Problems of overload, fatigue failure, wear and poor lubrication can be differentiated from the appearance of the teeth.

Vibration Analysis

Modern condition monitoring techniques encompass many different themes; one of the most important and informative is the vibration analysis of rotating machinery. Using vibration analysis, the state of a machine can be constantly monitored and detailed analysis may be made concerning the health of the machine and any faults which may arise or have already arisen. Machinery distress very often manifests itself in vibration or a change in vibration pattern. Vibration analysis is therefore, a powerful diagnostic and troubleshooting tool of major process machinery.

On-load monitoring can be performed mainly in the following three ways.

- Periodic field measurements with portable instruments; this method provides information about long-term changes in the condition of plant. The portable instruments are employed with a high load factor and can often be placed in the care of only one man. Use of life curves and the LEO approach assist the decision making.

- Continuous monitoring with permanently installed instruments; it is employed when machine failures are known to occur rapidly and when the results of such failure are totally unacceptable as in the case of turbine generator units.

- Signature analysis: scientific collection of information, signals or signatures, diagnosis and detection of the faults by a thorough analysis of these signatures based on the knowledge hitherto acquired in the field, and judging the severity of faults for decision making, all put together, is called signature analysis. The technique involves the use of electronic instrumentation especially designed for the purpose of varied capacities, modes of application and design features.

Vibration signals are the most versatile parameters in machine condition monitoring techniques. Periodic vibration checks reveal whether troubles are present or impending. Vibration signature analysis reveals which part of the machine is defective and why. Although a number of vibration analysis techniques have been developed for this purpose, still a lot of scope is there to reach a stage of expertise.

Temperature Monitoring

Temperature monitoring consists of measuring of the operational temperature and the temperature of component surfaces. Monitoring operational temperature can be considered as a subset of the operational variables for performance monitoring. The monitoring of component temperature has been found to relate to wear occurring in machine elements, particularly in journal bearings, where lubrication is either inadequate or absent. The techniques for monitoring temperature of machine components can include the use of optical pyrometers, thermocouples, thermography, and resistance thermometers.

Acoustic Emission Analysis

Acoustic emission refers to the generation of transient waves during the rapid release of energy from localized sources within a material. The source of these emissions is closely associated with the dislocation accompanying plastic deformation and the initiation or extension of fatigue cracks in material under stress. The other sources of acoustic emission are melting, phase transformations, thermal stress, cool-down cracking, and the failure of bonds and fibers in composite materials. Acoustic emissions are measured by piezoelectric transducers mounted on the surface of the structure under test and loading the structure. Sensors are coupled to the structure by means of a fluid couplant or by adhesive bonds. The output of each piezoelectric sensor is amplified through a low-noise preamplifier, filtered to remove any extraneous noise and furthered processed by suitable electronic equipment.

Traditionally, acoustic emissions as a technique has been restricted to the monitoring of high cost structures due to the expenses of the monitoring equipment. However, as equipment costs steadily fall, the range of viable applications expands rapidly. Olsson et al. present a frame work for fault diagnosis of industrial robots using acoustic signals and case-based reasoning [12]. This frame work utilizes the case-based reasoning for fault identification based on sound recording in robot fault diagnosis. Wue et al. have developed experimental setup for online fault detection and analysis of modern water hydraulic system [13], and suggested that the incorporation of wavelet transformation into the analysis of acoustic emission opens up the door for future research, which can prove to be very relevant toward condition monitoring. Choe et al. [14] worked on neural pattern identification of railroad wheel-bearing faults from audible acoustic signals by comparison of FFT, continuous wavelets transform (CWT) and discrete wavelets transform (DWT) features.

Noise Analysis

Noise signals are utilized for condition monitoring because noise signals measured at regions in proximity to the external surface of machines can contain vital information about the internal processes, and can provide valuable information about a machine's running condition. When machines are in a good condition, their noise frequency spectra have characteristic shapes. As faults begin to develop, the frequency spectra change. Each component in the frequency spectrum can be related to a specific source within the machine. This is the fundamental basis for using noise measurement and analysis in condition monitoring. Sometimes the signal which is to be monitored is submerged within some other signal and it cannot be detected by a straightforward time history or spectral analysis. In this case, specialized signal processing techniques have to be utilized.

Wear Debris Analysis

It is not possible to examine the working parts of a complex machine on load, nor is it convenient to strip down the machine. However, the oil which circulates through the machine carries with it evidence of the condition of parts encountered. Examination of the oil, any particle it has carried with it, allows monitoring of the machine on load or at shutdown. A number of techniques are applied, some very simple, other involving painstaking tests and expensive equipments. Presently, available lubricant sampling or monitoring techniques like rotary particles depositor (RPD), spectrophotometer oil analysis programme (SOAP), Ferrographic oil analysis and recent software used techniques are available to distinguish between damage debris and normal wear debris. Every machine ever designed undergoes a process of wear and tear in operation, yet a battery of modern condition monitoring techniques is available to monitor this process and trigger preventive maintenance routines which depend on identifying any problem before it has the chance to develop to the point of final breakdown. Now recently, engineers have been able to extend their knowledge of conditions within operating machinery by studying the particles of metallic debris which can be found in lubricating oil from engines, gearboxes, final drive units and transmissions, or in hydraulic fluid, and recording the number, size, and type of these fragments of debris.

Motor Current Signature Analysis

Motor current signature analysis (MCSA) is a novel diagnostic process for condition monitoring of electric motor-driven mechanical equipment (pumps, motor-operated valves, compressors, and processing machinery). The MCSA process identifies, characterizes, and trends overtime the instantaneous load variations of mechanical equipment in order to diagnose changes in the condition of the equipment. It monitors the instantaneous variations (noise content) in the electric current flowing through the power leads to the electric motor that drives the equipment. The motor itself thereby acts as a transducer, sensing large and small, long-term and rapid, mechanical load variations, and converting them to variations in the induced current generated in the motor windings. This motor current noise signature is detected, amplified, and further processed as needed to examine its time-domain and frequency-domain (spectral) characteristics. Korde [15] demonstrates that the spectrum analysis of the motors current and voltage signals can hence detect various faults without disturbing its operation using FFT transformation.

Nondestructive Testing

The principle of nondestructive testing (NDT) is to be able to use the components or structure after examination. The inspection should not affect the item involved, and must therefore, be nondestructive. NDT includes many different technologies, each suitable for one or more specific inspection tasks, with many different disciplines overlapping or complimenting others. Thus the best technique(s), for any one application, should be decided by an expert eddy current testing, electrical resistance testing, flux leakage testing, magnetic testing, penetrant testing, radiographic testing, resonant testing, thermographic testing, ultrasonic testing, and visual testing are some of the different NDT techniques.

VIBRATION SIGNATURE ANALYSIS

The word signature has been coined to designate signal patterns which characterize the state or condition of a system from which they are acquired. Signatures are extensively used as a diagnostic tool for mechanical system. In many cases, some kind of signal processing is undertaken on those signals in order to enhance or extract specific features of such vibration signatures. It is very important to consider the type and range of transducers used as pickup for capturing vibration signal. Signature-based diagnostic makes extensive use of signal processing techniques involving one or more methods to deal with the problem of improvement in the signal to noise ratio.

Vibration-based monitoring techniques have been widely used for detection and diagnosis of bearing defects for several decades. These methods have traditionally been applied, separately in time and frequency domains. A time-domain analysis focuses principally on statistical characteristics of vibration signal such as peak level, standard deviation, skewness, kurtosis, and crest factor. A frequency domain approach uses Fourier methods to transform the time-domain signal to the frequency domain, where further analysis is carried out, and conventionally using vibration amplitude and power spectra. It should be noted that use of either domain implicitly excludes the direct use of information present in the other. These techniques have been broadly classified in three areas, namely, the following.

Time-Domain Analysis

The time domain refers to a display or analysis of the vibration data as a function of time. The principal advantage of this format is that little or no data are lost prior to inspection. This allows for a great deal of detailed analysis. However, the disadvantage is that there is often too much data for easy and clear fault diagnosis. Time-domain analysis of vibration signals can be subdivided into the following categories: time-waveform analysis, time-waveform indices, time-synchronous averaging, negative averaging, orbits, and probability density moments.

Frequency Domain

The frequency domain refers to a display or analysis of the vibration data as a function of frequency. The time-domain vibration signal is typically processed into the frequency domain by applying a

Fourier transform, usually in the form of a fast Fourier transform (FFT) algorithm. The principal advantage of this format is that the repetitive nature of the vibration signal is clearly displayed as peaks in the frequency spectrum at the frequencies where the repetition takes place. This allows for faults, which usually generate specific characteristic frequency responses, to be detected early, diagnosed accurately, and trended overtime as the condition deteriorates. However, the disadvantage of frequency-domain analysis is that a significant amount of information (transients, nonrepetitive signal components) may be lost during the transformation process. This information is nonretrievable unless a permanent record of the raw vibration signal has been made. The various methods of frequency-domain vibration signature analysis are bandpass analysis, shock pulse (spike energy), enveloped spectrum, signature spectrum, and cascades (waterfall plots).

The Quefrency Domain

The quefrency is the abscissa for the cepstrum which is defined as the spectrum of the logarithm of the power spectrum. It is used to highlight periodicities that occur in the spectrum in the same manner as the spectrum is used to highlight periodic components occurring in the time domain [16]. One of the ways the expert system detects bearing tones is by looking at the spectrum of a spectrum. This process is called cepstrum analysis, "cepstrum" being a play on the word "spectrum."

FAULT DETECTION FROM VIBRATION ANALYSIS

Renwick and Babson [1] demonstrate that the predictive maintenance using vibration analysis has achieved meaningful results in successfully diagnosis machinery problems. The benefits of such programs include not only evident-cost benefits such as reducing machinery downtime and production losses, but also the more subtle long-term cost benefits which can result from accurate maintenance scheduling.

source identification and fault detection from vibration signals associated with items which involve rotational motion such as gears, rotors and shafts, rolling element bearings, journal bearings, flexible couplings, and electrical machines depend upon several factors: (i) the rotational speed of the items, (ii) the background noise and/or vibration level, (iii) the location of the monitoring transducer, (iv) the load sharing characteristics of the item, and (v) the dynamic interaction between the item and other items in contact with it.

The main causes of mechanical vibration are unbalance, misalignment, looseness and distortion, defective bearings, gearing and coupling in accuracies, critical speeds, various form of resonance, bad drive belts, reciprocating forces, aerodynamic or hydrodynamic forces, oil whirl, friction whirl, rotor/stator misalignments, bent rotor shafts, defective rotor bars, and so on. Some of the most common faults that can be detected using vibration analysis are summarized in Table 2 [17].

Table 2: Some typical faults and defects that can be detected with vibration analysis

Item	Fault
Gears	Tooth messing faults,
	misalignment,
	cracked and/or worm teeth,
	eccentric gear

Rotors and shaft	Unbalance
	Bent shaft
	Misalignment
	Eccentric journals
	Loose components
	Rubs
	Critical speed
	Cracked shaft
	Blade loss
	Blade resonance
Rolling element bearings	Pitting of race and ball/roller
	Spalling
	Other rolling elements defect
Journal bearing	Oil whirl
	Oval or barreled journal
	Journal/bearing rub
Flexible coupling	Misalignment
	Unbalance
Electrical machines	Unbalanced magnetic pulls
	Broken/damaged rotor bars
	Air gap geometry variations
	Structural and foundation faults
	Structural resonance
	Piping resonance
	Vortex shedding

Weqerich et al. developed a nonparametric modeling technique by smart signal and demonstrate the use of this approach for detecting faults in rotating machinery via extracted features from vibration signals [18]. Lei et al. [19] present damage diagnosis approach using time series analysis of vibration signals for structural health monitoring benchmark problem.

Sohn and Farrar [20] have presented a procedure for damage detection and localization within a mechanical system solely based on the time series analysis of vibration data. Sahinkaya et al. [21] have worked on fault detection and tolerance in synchronous vibration control of rotor magnetic bearing system. A simple and effective algorithm has been developed to build fault detection and tolerances capabilities into the open-loop adaptive control of the synchronous vibration of flexible rotors supported or equipped with magnetic bearings.

Lebold et al. [22] have presented review of vibration-analysis methods for gearbox diagnostics and prognostics. This review listed some of the most traditional features used for machinery diagnostics and presented some of the signal processing parameters that impact their sensitivity.

Verma and Balan [23] present a fundamental study on the vibration behavior of electrical machine stators using an experimental model analysis and suggested that vibration level even at resonance can be reduced by designing the electromagnetic forces to have circumferential mode associated with corresponding resonance. Ocak and Loparo [24] present algorithms for estimating the running speed and the bearing-key frequencies of an induction motor using vibration data that can be used for failure detection and diagnosis. Plenge et al. [25] developed optical inspection techniques for vibration analysis and defect indication in railway.

Vibration signals from gearboxes and roller bearings share many common characteristics. First, the signals are usually noisy. This is because the accelerometers for signals collection are mounted on the outer surface of gearbox. The signals obtained from these accelerometers include vibrations from meshing gears, bearings, and the equipment's many other parts. Second, symptoms from faulty bearings are very similar to those from faulty gears. For example, periodic impulses may indicate either cracked teeth of gears or damaged races or rollers of roller bearings. Such periodic impulses, however, cannot be detected easily with the frequency spectrum because of the heavy noise distributed in the low-frequency area [4].

Lin et al. [4] have obtained excellent results for mechanical fault detection based on the wavelet denoising technique. The method has performed excellently when used to denoise mechanical vibration signals with a low signal-to-noise ratio.

REB FAILURE AND ITS VIBRATION ANALYSIS

Each of the rolling element bearings used in industries consists of two rings, one inner and the other outer. A set of balls or rolling elements placed in raceways rotate inside these rings. Even when properly applied and maintained, the bearing will still be subjected to one cause of failure, fatigue of bearing material. Fatigue is the result of shear stresses cyclically applied immediately below the load-carrying surfaces and is observed as spalling away of surface metal. However, material fatigue is not the only cause of spalling. There are causes of premature spalling. So, although the observer can identify spalling, he must be able to discern between spalling produced at the normal end of bearing's useful life and that triggered by causes found in the three major classifications of premature spalling as lubrication, mechanical damage, and material defects. Most bearing failures can be attributed to one or more of the following causes as defective bearing seats on shafts and in housings, misalignment, faulty mounting practice, incorrect shaft and housing fit, inadequate lubrication, ineffective sealing, and vibration, while the bearing is not rotating, passage of electric current through the bearing [2]. As one of the most essential parts in rotating machinery rolling element bearings are often subjected to high stress and operate under severe conditions, Their integrity becomes an issue particularly in key machinery. A machine could be seriously jeopardized, if defects occur to those rolling element bearing during service [5]. A new approach for the categorization of bearing faults was introduced by Stack et al. [26]. A common way in which bearing faults are often classified according to the location of the fault (an inner race/outer race/ball/cage fault). In this research, bearing faults were grouped into one of two categories, as single point defects and generalized roughness. The single point defects are defined as visible defects that appear on the raceways, rolling elements, or cage. A single point defect produces one of the four characteristics fault frequencies depending on which surface of the bearing contains the fault. In spite of the name, a bearing can possess multiple single point defects. The other group of bearing faults, generalized roughness, refers to an unhealthy bearing whose damage is not apparent to the unaided eye. Example of this failure mode includes deformation or warping of the rolling elements or raceways and overall surface roughness due to heating, contaminated lubricant, or electric discharge machining. The effects produced by this failure mode are difficult to predict, and there are no characteristics fault frequencies with this type of fault.

In rolling element bearing failure analysis, the low-frequency phenomenon is the impact caused by a defect of a bearing. The high-frequency carrier is a combination of the natural frequencies of the associated rolling element or even of the machine [27]. There are a number of factors that contribute to the complexity of the bearing signature. First, variation of bearing geometry and assembly make it impossible to precisely determine bearing characteristics frequencies. Secondly, locations of bearing defects cause different behavior in the transient response of the signal, which is easily buried in wide band response and noisy signals. Thirdly, signature appears to be very different with the same type of defect at different stages of damage, severity. Finally, operating speed and loads of the shaft greatly affect the way and the amount a machine vibrates.

Several researchers worked on the subject of rolling element bearing defect detection and diagnosis through vibration analysis. Time domain, frequency domain, time-frequency domain based on short time Fourier transform (STFT) and wavelet transform and advanced signal processing techniques have been implemented and tested. Time-domain analysis focuses on dealing directly with the time-domain waveform of vibration signals. The indices RMS, peak value, and crest factor are often used to quantify the time signal. The statistical parameters such as kurtosis and skewness values are robust to varying bearing operating condition and are good indicators of incipient defects. The disadvantage, however, is that as the defect spreads across the bearing surfaces the values of these parameters drop back to normal [28].

The frequency domain, spectrum of the vibration signal reveals frequency characteristics of vibration. If the frequencies of the impulse occurrence are close to one of the bearing characteristic frequencies, such as ball pass inner race frequency, ball pass outer race frequency, ball spin frequencies, it may indicate a defect related fault in the bearing. Fast Fourier transform is used in conventional frequency-domain signature analysis techniques for conversion of time-domain signal in frequency-domain signal. Other frequency-domain techniques are generally used are the calculation of power spectral density, bandpass analysis, envelope analysis. The effectiveness of bandpass-analysis method relies on a suitable choice of narrow-band frequencies around the selected resonance. In envelope analysis, signals are filtered through bandpass filter and filtered signal is demodulated with the help of full-wave rectification or via Hilbert transform and then spectrum analyzed. The passband and envelope analysis techniques are useful to detect rolling element bearing faults when signals are noisy due to severity of fault or due to associated noise from other sources as shaft misalignment, unbalance, and looseness.

The fast Fourier transform has drawback, when signal is nonstationary or noisy, even in FFT, time information is lost. Many researchers have used short-time Fourier transform (STFT) to overcome the time information problem but low-resolution problem exists in STFT. The wavelets transform is currently used to overcome both the time information and low resolution problems. A major advantage of the wavelets transform is that this method can exhibit the local features of the signals and give account of how energy is distributed over frequencies changes from one instant to the next.

The confidence of bearing fault diagnosis can be improved by using a range of failure indicators including performance indices, oil analysis, thermography, and motor current readings in conjunction with vibration analysis. These indicators are generally assimilated and analyzed by human expert but a computational expert system, based on neural network, fuzzy logic, and rule based logic, as well as hybrid techniques containing elements of all three methods, is being used and continually improved in order to automate the process.

The neural network technology provides an attractive complement to traditional vibration analysis because of the potential of neural networks to operate in real-time mode and to handle data that may be distorted or noisy [29]. Neural networks have proven the ability in the area of nonlinear pattern classification and can correctly identify the different causes of bearing vibration [30]. The fuzzy logic has proven ability in mimicking human decisions, and the bearing fault diagnosis problem has typically been solved by an experienced engineer. The fuzzy logic is promising for automation in the area of bearing vibration diagnosis, if the input data is well processed [31]. The advantages of the fuzzy logic approach include the possibility to change the linguistic rules into decisions by copying the procedure and thinking of a human analyzer. The rules that include uncertainty and inaccuracy are changed into numbers describing the severity or the probability of fault. The rules and membership functions can be tuned to find the good sensitivity of the diagnostic system.

ADVANCED SIGNAL PROCESSING TECHNIQUES IN VIBRATION ANALYSIS

With the development of soft computing techniques such as artificial neural network (ANN) and fuzzy logic, there is a growing interest in applying these approaches to the different areas of engineering. These systems gained popularity over other methods, as they are model free estimators capable of synthesizing nonlinear and noisy systems. The fuzzy logic was developed as a mean for representing, manipulating, and utilizing uncertain information (information that is usually expressed in linguistic terms). The recent surge of interest is in merging or combining NN and fuzzy logic system into a functional system to overcome their individual weaknesses.

Wavelet analysis is an emerging field of mathematics that has provided new tool and algorithms for the type of problems encountered in process monitoring. Wavelet transform (WT) is a mathematical approach that decomposes a time-domain signal into different frequency groups. Wavelet algorithms process data at different scales and resolutions

The monitoring and diagnosis of machinery is a well-established discipline, but much progress remains to be made in automating diagnosis as well as developing low-cost reliable technologies which can be applied cost-effectively in the majority of production environment. Developments in microtechnology and artificial intelligence have driven the trends toward more extensive onboard diagnostics. Recent systems have relied on artificial intelligence techniques to strengthen the robustness of diagnostics systems. Four artificial techniques have been widely applied as expert system, neural networks, fuzzy logic, and model-based systems [9]. Different kinds of artificial intelligence method have become common in fault diagnosis and condition monitoring. For example, fuzzy logic and neural networks have been used in modeling and decision making in diagnostics schemes. Neural networks-based classifications are used in diagnosis of rolling element bearings.

Shikari and Sadiwala worked on automation in condition-based maintenance using vibration analysis [32]. In this work, importance of intelligent system in CBM is focused. Dyke [33] describes an example of the application of the DLI engineering ExpertAlert expert automated diagnostics system to successful diagnosis of machine tool spindle bearing problems. Sima [34] proposes a strictly neural expert system architecture that enables the creation of the knowledge base automatically by learning from example inferences. Bandyopadhya et al. [35] have developed an expert system for real-time condition monitoring using vibration analysis for turbine bearing. Poyhonen et al. [36] have applied support vector classification to fault diagnostics of an electrical machine.

Zhenya et al. proposed a multilayer feed forward network-based machine state identification method. They represent certain fuzzy relationship between the fault symptoms and causes, with highly nonlinearity between the input and the output of the network [37]. The rolling element bearing signals are investigated accordingly to the principle that the wavelet can extract the signal envelope by Jun and Liao. A wavelet-based self-information extracting envelope method was applied, application of the method demonstrates that the method is effective to extract the rolling bearing signal envelope and is useful to analysis the bearing faults [27].

Four approaches based on bispectral and wavelet analysis of vibration signals are investigated as signal processing techniques for application of a number of induction motor rolling element bearing faults by Yang et al. [38]. A general methodology for machinery fault diagnosis through a pattern recognition technique is developed by Sun et al. [28], this involves data acquisition, feature extraction, mapping for feature fusion, and piecewise-linear classification and diagnosis. They conclude, to increase the sensitivity and reliability of pattern recognition, one is encouraged to include as many feature parameters as possible without concern the redundancy or numerical singularities.

Satish and Sharma [39] demonstrate a novel and cost-effective approach for diagnosis and prognosis of bearing faults in small and medium-size induction motor. In this work, a fuzzy back-propagation network was developed by combining neural network with fuzzy logic to identify the present condition of the bearing and to estimate the remaining life of the motor.

Fan and Zuo [40] proposed an effective method to extract modulating signal and to detect the early gear fault. In this new fault detection method, combination of Hilbert transform and wavelet-packet transform were used. Both simulated signals and real vibration signals collected from a gearbox dynamics simulator were used to verify the proposed method

Duraisamy et al. [41] have described a comparative study of membership functions for design of fuzzy logic fault diagnosis system for single-phase induction motor.

Intelligent systems cover a wide range of techniques related to hard science such as modeling and control theory, and soft science such as the artificial intelligence. Intelligent systems, including neural networks, fuzzy logic, and wavelet techniques, utilize the concepts of biological systems and human cognitive capabilities. These three systems have been recognized as a robust and alternative to some of the classical modeling and control methods [24].

CONCLUSIONS

In this paper, authors have been presented a brief review of art of machinery fault detection, different conventional and recent techniques were discussed for machine fault signature analysis with particular regard to rolling contact bearing fault diagnosis through vibration analysis. After the review of literature on machine fault signature analysis, the following points are concluded.

- Prevention of potential failure is required for reliable and safe operations of machineries and the prevention of catastrophic failure can be done by appropriate maintenance. Condition-based maintenance is the best suitable technique to avoid unwanted futuristic failures through condition monitoring or signature analysis for rotating machineries. Vibration signature analysis is the best suitable technique available for fault identification

- Among all machine components rolling contact bearing is needed more attention towards signature analysis. The lot of scope is available in bearing fault signature analysis through vibration data for multiple points or generalized faults.

- Vibration analog signal can be converted in discrete data for further investigation and various time-domain and frequency-domain features can be used for further investigations. The Hilbert and wavelets transform have tremendous scope in machine fault signature analysis.

- Expert system based on ANN and fuzzy logic can be developed for robust fault categorization with the use of extracted features from vibration signal.

These conclusions motivate further research to incorporate other parameters and symptoms with vibration features to develop more robust expert systems for machine faults signature analysis.

ACKNOWLEDGMENTS

The authors are pleased to acknowledge the support of Madhav Institute of Technology and Science (MITS), Gwalior, India, for providing the facility of literature review. A special thanks to Dr. S. Wadhwani, Lecturer, Electrical Engineering Department, MITS Gwalior, India, for the motivation throughout the literature survey.

REFERENCES

1. J. T. Renwick and P. E. Babson, "Vibration analysis-a proven techniques as a productions maintenance tool," IEEE Transactions on Industry Applications, vol. 21, no. 2, pp. 324–332, 1985.

2. H. P. Bloch and F. K. Geitner, in Machinery Failure Analysis and Troubleshooting, chapter 5, Gulf Publishing Company, Houston, Tex, USA, 1983.

3. M. Amarnath, R. Shrinidhi, A. Ramachandra, and S. B. Kandagal, "Prediction of defects in antifriction bearings using vibration signal analysis," Journal of the Institution of Engineers India Part MC Mechanical Engineering Division, vol. 85, no. 2, pp. 88–92, 2004.

4. J. Lin, M. J. Zuo, and K. R. Fyfe, "Mechanical fault detection based on the wavelet de-noising technique,"Journal of Vibration and Acoustics, vol. 126, no. 1, pp. 9–16, 2004.

5. D. F. Shi, W. J. Wang, and L. S. Qu, "Defect detection for bearings using envelope spectra of wavelet transform," Journal of Vibration and Acoustics, vol. 126, no. 4, pp. 567–573, 2004.

6. S. Edwards, A. W. Lees, and M. I. Friswell, "Fault diagnosis of rotating machinery," The Shock and Vibration Digest, vol. 30, no. 1, pp. 4–13, 1998.

7. M. Dileo, C. Manker, Cadick, and P. E. Jhon, "Condition based maintenance," October 1999, Cadick Corporation.

8.   A. K. Gupta, in Reliability Engineering and Terotechnology, chapter 13, Macmillan India, New Delhi, India, 1996.

9.   L. Mann, A. Saxena, and G. M. Knapp, "Statistical-based or condition-based preventive maintenance?"Journal of Quality in Maintenance Engineering, vol. 1, no. 1, pp. 1355–2511, 1995.

10.   S. Nandi and H. A. Toliyat, "Condition monitoring and fault diagnosis of electrical machines-a review," in Proceedings of the 34th IEEE IAS Annual Meeting on Industry Applications Conference, vol. 1, pp. 197–204, Phoenix, Ariz, USA, October 1999.

11.   B. Marcus, "Condition based maintenance on Rail vehicle— possibilities for an innovation," 2002, Design and Product development, Nalandalen University, Eskilsting Sweden.

12.   E. Olsson, P. Funk, and M. Bengtsson, "Fault diagnosis of industrial robots using acoustic signals and case-based reasoning," in Proceedings of the 7th European Conference on Case Based Reasoning (ECCBR ‹04), vol. 3155 of Lecture Notes in Computer Science, pp. 686–701, Springer, Madrid, Spain, August-September 2004.

13.   T. K. Wue, P. S. K. Chua, and G. H. Lim, "On line fault detection and analysis of modern water hydraulic system," Journal of Institution of Engineers, vol. 44, no. 4, pp. 41–52, 2004.

14.   H. C. Choe, Y. Wan, and A. K. Chan, "Neural pattern identification of railroad wheel-beating faults from audible acoustic signals: comparison of cost," CWT and DWT features, Department of Electrical Engineering, Taxes A&M University.

15.   A. Korde, "On line condition monitoring of motors using electrical signature analysis," in Proceedings of the 4th International Conference on Engineering and Automation, Orlando, Florida, USA, July-August 2000.

16.   in Handbook of Condition Monitoring Techniques and Methodology, A. Davies, Ed., chapter 12, Chapman and Hall, London, UK, 1998.

17. M. P. Norton and D. G. Karczub, in Fundamentals of Noise and Vibration Analysis for Engineers, chapter 8, United Kingdom at the University Press, Cambridge, UK, 2nd edition, 2003.

18. S. W. Weqerich, A. D. Wilks, and R. M. Pipke, "Nonparametric modeling of vibration signal features for equipment health monitoring," in Proceedings of the IEEE Aerospace Conference, vol. 7, pp. 3113–3121, Big Sky, Mont, USA, March 2003.

19. Y. Lei, M. Kire, and T. W. Kenny, "Statistical damage detection using time series analysis on a structural health monitoring benchmark problem," in Proceedings of the 9th International Conference on Applications of Statistics and Probability in Civil Engineering, San Francisco, Calif, USA, July 2003.

20. H. Sohn and C. R. Farrar, "Damage diagnosis using time series analysis of vibration signals," Journal of Smart Material and Structures Institute of Physics, vol. 10, pp. 446–451, 2001.

21. M. N. Sahinkaya, M. O. T. Cole, and C. R. Burrows, "Fault detection and tolerance in synchronous vibration control of rotor-magnetic bearing system," Proceedings of the Institution of Mechanical Engineers, Part C: Journal of Mechanical Engineering Science, vol. 215, no. 12, pp. 1401–1416, 2001.

22. M. Lebold, K. McClintic, R. Campbell, C. Byington, and K. Maynard, "Review of vibration analysis method for gearbox diagnostic and prognostics," in Proceedings of the 54th Meeting of the Society for Machinery Failure Prevention Technology, pp. 623–634, Virginia Beach, Va, USA, May 2000.

23. S. P. Verma and A. Balan, "Experimental investigations on the stations of electrical machines in relation to vibration and noise problems," IEE Proceedings Electric Power Applications, vol. 145, no. 5, pp. 455–461, 1998.

24. H. Ocak and K. A. Loparo, "Estimating the running speed and the bearing key frequencies of an instruction motor from vibration data," Mechanical Systems and Signal Processing, vol. 18, no. 3, pp. 515–533, 2004.

25. M. Plenge, R. Lammering, T. Walz, and A. Ettemeyer, "Optical

inspection techniques for vibration analysis and defect indication in railway," in Proceedings of the 16th WCNDT World Conference on NDT, Montreal, Canada, August-September 2004.

26.   J. R. Stack, T. G. Habetler, and R. G. Harley, "Fault classification and fault signature production for rolling element bearings," in Proceeding of the 4th IEEE International Symposium on Diagnostics for Electric Machines, Power Electronics and Drives (SDEMPED ‹03), pp. 172–176, Atlanta, Ga, USA, August 2003.

27.   P. Jun and M. F. Lias, "A wavelet based method for extracting rolling element bearing vibration signal envelope," in Proceeding of the WSEAS International Conference on Computer Engineering and Applications, pp. 181–186, Gold Coast, Australia, January 2007.

28.   Q. Sun, P. Chen, D. Zhang, and F. Xi, "Pattern recognition for automatic machinery fault diagnosis," Journal of Vibration and Acoustics, vol. 126, no. 2, pp. 307–316, 2004.

29.   I. E. Alguindigue, A. E. Wicz-Buczak, and R. E. Uhrig, "Monitoring and diagnosis of rolling element bearings using artificial neural networks," IEEE Transactions on Industrial Electronics, vol. 40, no. 2, pp. 209–217, 1993.

30.   B. Li, G. Goddu, and M. Y. Chow, "Detection of common motor bearing faults using frequency-domain vibration signals and a neural network based approach," in Proceeding of the American Control Conference, vol. 4, pp. 2032–2036, Philadelphia, Pa, USA, June 1998.

31.   G. Goddu, B. Li, M.-Y. Chow, and J. C. Hung, "Motor bearing fault diagnosis by a fundamental frequency amplitude based fuzzy decision system," in Proceedings of the 24th Annual Conference of the IEEE Industrial Electronics Society (IECON ‹98), vol. 4, pp. 1961–1965, Aachen, Germany, August-September 1998.

32.   B. Shikari and C. M. Sadiwala, "Automation in Condition Based Maintenance using vibration analysis," 2004, MANIT Bhopal.

33. J. V. P .E. Dyke, "Using an expert system for precision machine tool diagnostics: a case study," 1998, DLI engineering.

34. J. Sima, "Neural expert systems," Neural Networks, vol. 8, no. 2, pp. 261–271, 1995.

35. A. Badyopadhya, S. D. Mandal, and B. Pal, "Real time condition monitoring system using vibration analysis for turbine bearing," speech and signal processing Group Calcutta, India.

36. S. Poyhonen, M. Negrea, A. Arkkio, H. Hyotyniemi, and H. Koivo, "Support vector classification for fault diagnosis of an electrical machine," in Proceedings of the 6th International Conference on Signal Processing (ICSP ‹02), vol. 2, pp. 1719–1722, Helsinki University of Technology, Finland, August 2002.

37. H. Zhenya, W. Meng, and G. Bi, "Neural network and its application on machinery fault diagnosis," inProceedings of the IEEE International Conference on Systems Engineering, pp. 576–579, Kobe, Japan, September 1992.

38. D. M. Yang, A. F. Stronach, and P. MacConnell, "The application of advanced signal processing techniques to induction motor bearing condition monitoring," Meccanica, vol. 38, no. 2, pp. 297–308, 2003.

39. B. Satish and D. R. Sharma, "A fuzzy BP approach for diagnostics and prognosis of bearing faults in induction motors," in IEEE Power Engineering Society General Meeting, vol. 3, pp. 2291–2294, San Francisco, Calif, USA, June 2005.

40. X. Fan and M. J. Zuo, "Gearbox fault detection using Hilbert and wavelet packet transform," Mechanical Systems and Signal Processing, vol. 20, no. 4, pp. 966–982, 2006.

41. V. Duraisamy, N. Devarajan, and D. Somasundareswari, "Comparative study of membership functions for design of fuzzy logic fault diagnosis system for single phase induction motor," Academic Open Internet Journal, vol. 13, pp. 1–6, 2004.

Chapter 2

Failure Analysis of Belt/Roll Tribological Pair used for the Production of Eucalypt Fiber Panels

Cesar R.F. Azevedo[a], J.C.K. das Neves[b], and A. Sinatora[c]

[a]Department of Metallurgical and Materials Engineering, Escola Politécnica da Universidade de São Paulo, Brazil

[b]Mechanical Department, Universidade Tecnológica Federal do Paraná, Brazil

[c]Surface Phenomena Laboratory, Mechanical Engineering Department, Escola Politécnica da Universidade de São Paulo, Brazil

ABSTRACT

The premature failure of a large agglomeration machine used for the annual production of 360,000 m³ of eucalypt fiber panels was investigated to identify the nucleation and growth mechanisms of cracking in PH stainless steel belts (126 m × 2.9 m × 3.0 mm). These belts are used to compress a cushion composed of eucalyptus fibers and glue, being the pressure transmitted from the pistons by the action of numerous case-hardening steel rolls. Examination of the belt working interfaces (belt/rolls and belt/eucalypt fibers) indicated that the main cracking was nucleated on the belt/roll interface and that there is a clear relationship between the crack nucleation and the presence of superficial irregularities, which were observed on the belt/roll working surface. Used rolls showed the presence of perimetric wear marks and 2 μm silicon-rich encrusted particles (identified as silicon carbide). Lubricant residues contained the presence of helicoidal wires, which were originated by the release of the stainless steel cleaning brush bristles, and 15 μm diameter metallic particles, which were generated by material detachment of the belt. The presence of foreign particles on the tribological interface contributed to an increase of the shear stresses at the surfaces and, consequently, the number of the contact fatigue crack nucleation sites in the belt/roll tribo-interface. The cracking was originated on the belt/roll interface of the stainless steel belt by a mixed rolling/slip contact fatigue mechanism, which promoted spalling and further nucleation and growth of conventional fatigue cracks. Finally, the system lubrication efficiency and the cleaning procedure should be optimised in order to increase the life expectancy of the belt.

INTRODUCTION

Countries like New Zealand, Chile, Brazil, and South Africa have large, intensively managed plantations of exotic species, where many softwood plantations are being replanted with fast growing hardwoods (Eucalyptus). As more forests are set aside (reserved) in developed nations and more deforestation occurs in developing nations, industrial plantations will be ever more critical in meeting the worlds need for industrial wood fiber [1]. Implementing an industrial forest has consequences on social, economic and ecological levels and large-scale forestry projects, especially tree plantation, must avoid usurpation of agricultural lands, replacement of valuable native ecosystems, depletion of water resources, worsening inequity in land ownership and evictions of local population [1], [2] and [3].

The wood, paper and cellulose industry in Brazil employs almost 2 million workers, generating a revenue of US$ 21 billion/year, which corresponds to 4% of the Brazilian Gross Internal Product (GIP). Additionally, this sector is responsible for 10% of the Brazilian exports. The forests in Brazil cover approximately 5.5 million km^2, accounting for more than 25% of the world's tropical forests. According to a former Brazilian environmental minister, there are over 1 million km^2 of deforested areas in Brazil without any economical usage. From the annual internal consumption of 300 million m^3 of wood in Brazil, only 1/3 comes from industrial forests, which account for an area of just 50.000 km^2[1], [2] and [3]. In this context, companies that use raw material from well-managed industrial forests play an important role not only in social and economic terms but also in terms of rational use of natural resources. Taking a look to the productive process of these companies it is found that their machinery usually works under severe conditions, especially if it was considered that fiber wood is an intrinsically heterogeneous raw material.

The present paper will investigate the premature failure of a large agglomeration machine used for the annual production of 360,000 m^3 of eucalypt fiber panels. This failure was caused by the presence of through-thickness cracks in the PH stainless steel belts (126 m × 2.9 m × 3.0 mm). These belts are used to compress a cushion composed of eucalyptus fibers and glue, being the pressure transmitted from the pistons by the action of numerous case-hardening steel rolls. The cracks were observed just after 19 months of operation and they were preferentially located on the central part of the upper belt (the expected life should be around 60 months). After the detection of the transversal through-thickness crack in the stainless steel belt, a maintenance halt was carried out. It consists of the crack welding via TIG (tungsten inert gas) process, followed by grinding and polishing of the welded surfaces. In case of a relapsing cracking in the belt, an area of 30 cm diameter containing the crack is removed from the belt (by plasma cutting) and replaced by a new stainless steel disc by a sequence of welding, grinding and polishing operations. The agglomeration press contains two stainless belts (126 m × 2.9 m × 3.0 mm) moving at a maximum velocity of 90 m/min and positioned symmetrically to "insandwich" a cushion composed of eucalyptus fibers and glue, producing, as a result, a eucalyptus fiber panel with density between 650 and 1100 kg/m³. The contact temperature between the belts and eucalyptus cushion is about 250 °C. The belt/eucalyptus cushion contact area is around (50 m × 2.90 m) – see Fig. 1. The agglomeration press is composed by 82 frames, each presenting five hydraulic pistons, amounting 410 pistons. The pressure is transmitted from the pistons to the belt by the action of numerous case-hardening steel rolls (supplied in the polished condition, with approximately 2.9 m length, 18 mm diameter and assembled with a 20 mm inter-roll spacing). The belt/roll interface is lubricated by means of a synthetic lubricant spray. The pressure varies with the product's thickness and it is controlled by the hydraulic pistons positioned along the belt's width and length (project's pressure = 500 N/cm², nominal pressure = 470 N/cm²). The working pressure is kept high and constant along the first 20 m of the agglomeration press, being then reduced to a value between 150 and 300 N/cm².

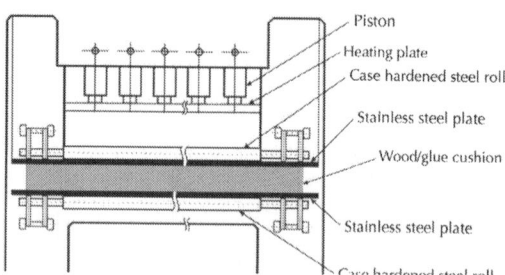

Figure 1: Transversal section, scheme showing the functioning of the agglomeration press, featuring the positioning of rolls, stainless steel belts and wood/glue cushion.

EXPERIMENTAL PROCEDURE AND RESULTS

The present investigation characterised new and cracked belts, new and used rolls and lubrication residues found next to the roll/belt interface in order to identify the mechanism of formation and growth of a transversal through-thickness cracking found in the PH stainless steel belts. Two stainless steel belt segments (new and used) were made available for the current investigation and the used belt presented a transversal through-thickness crack, which led to a maintenance halt of the agglomeration press. Additionally, three case hardening steel rolls (one new and two used: upper and lower) and three types of residues were also made available for the investigation. The studied samples are shown in Table 1 and the following tests were carried out: chemical analysis, tensile testing, visual inspection, macro and microstructural characterisation, topographic and microfractographic examination.

Table 1: List of samples

Sample	Description
1	New belt
2	Used upper belt (crack)
3	Roll 1 (new)
4	Roll 2 (used – upper)
5	Roll 3 (used – lower)
6	Dry residue (wood)
7	Pasty residue (wood + lubricant)
8	Liquid residue (burn out lubricant)

Chemical Analysis and Tensile Testing

Chemical analysis results indicated that the belt material is a PH (precipitation hardened) stainless steel, with chemical composition close to Custom 450 (XM-25) steel, except for silicon, chromium and niobium values (see Table 2). The chemical composition of the case-hardening steel roll is in accordance with the requirements of a SAE 1055 carbon steel. Tensile testing results (see Table 3) indicated that the belt's properties are compatible with the values of a Custom 450 (XM-25) stainless steel plate heat-treated for H900 condition. The ultimate tensile strength value of the used belt is slightly lower than the new belt and the elongation values for both belts are almost 100% higher that the value listed in the material certificate.

Table 2: Stainless steel belt – chemical analysis results

Element	Used belt	New belt	Typical values	Custom 450 (XM-25)[a]
Carbon (C)	0.04 ± 0.01	0.04 ± 0.01	0.04	0.05 max.
Silicon (Si)	1.56 ± 0.03	1.52 ± 0.03	1.52	1.00 max.
Manganese (Mn)	0.34 ± 0.01	0.34 ± 0.01	0.28	1.00 max.
Phosphor (P)	0.022 ± 0.002	0.021 ± 0.002	0.028	0.030
Sulphur (S)	<0.004	<0.004	0.001	0.030
Chromium(Cr)	13.7 ± 0.1	13.8 ± 0.1	13.71	14.0–16.0
Nickel (Ni)	7.1 ± 0.2	7.0 ± 0.1	6.91	5.0–7.0
Molybdenum (Mo)	0.76 ± 0.02	0.75 ± 0.01	–	0.5–1.0 Mo
Copper (Cu)	1.4 ± 0.1	1.3 ± 0.1	0.7	1.25–1.75
Titanium (Ti)	0.37 ± 0.02	0.34 ± 0.01	0.35	Nb min 0.4

[a]ASM Handbook, vol. 1, Wrought stainless steel, p. 841–907.

Table 3: Stainless steel belt – tensile testing results

Parameter	Used belt	New belt	Typical values	Custom 450 H900[a]
Ultimate tensile strength (MPa)	1451 ± 9	1527 ± 9	1450–1550	1240 min.
Yield strength (MPa)	n.d.	n.d.	1410–1500	1170 min.

| Elongation (%) | 11.3 ± 0.7 | 10.4 ± 0.7 | 6–8 | 4 min. |
| Hardness (HV) | 480[b] | 510[b] | 460–480 | 382 min. |

[a]ASM Handbook, vol. 1, Wrought stainless steel, p. 841–907.
[b]Measurements taken on the roll/belt surface.

Visual Inspection and Microtopographic Examination

Rolls

Fig. 2a and b show the general aspects of the as-received rolls. The used rolls present a brown coloured surface and perimetric bright marks along its length. The extremities of the used rolls presented internal holes and the working surfaces contained blue-coloured stripes. The diameter of the rolls was measured along 10 points using a micrometer and the results did not revealed any macroscopic wear of the rolls. The surface of the new roll presents numerous perimetric marks due to surface finishing, which became fainted in the used condition (see Fig. 2c and d). The microtopographic features of the used rolls are shown in Fig. 3a–d. The used rolls generally presents an oxidised region with few fainted finishing marks and additional presence of perimetric marks, especially on central region and near the roll's extremities (see Fig. 3a and b). Further investigation showed that the region with perimetric marks is composed of indentation marks, cracking and presence of encrusted 2 µm Si-rich particles (see Fig. 3d). These Si-rich particles were also observed in some areas the oxidised region (see Fig. 3c). Energy dispersive spectroscopy (EDS) microanalyses were obtained using a 30° sample tilt; accelerating voltage of 15 kV and automatic standardless quantification using ZAF correction.

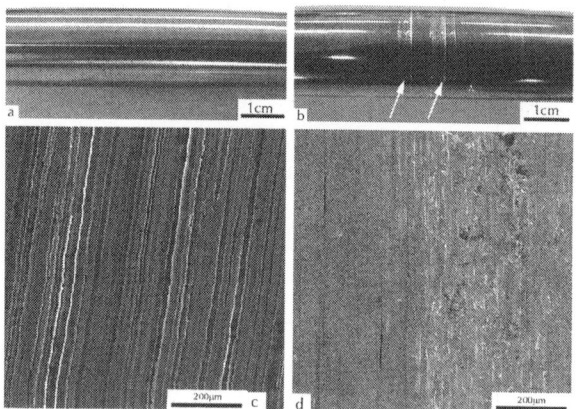

Figure 2: Rolls – (a) General view: new roll surface; (b) General view: used upper roll surface showing perimetric marks (see arrow) and superficial oxidation; (c) New roll: detail showing surface finishing marks, SEM-SEI; (d) Used roll: detail showing oxidised region with fainted finishing marks (left) and perimetric wear marks (right), SEM-SEI.

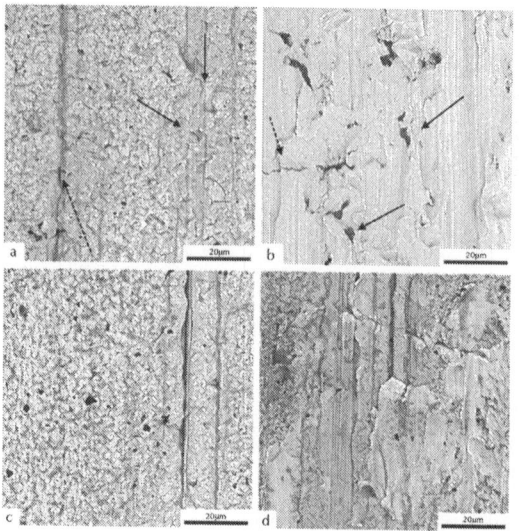

Figure 3: Rolls – microtopographic examination – (a) Used upper roll: detail of the oxidised region showing fainted surface finishing marks (dot-

ted arrow) and few indentation marks (arrow); (b) Used upper roll – detail of the regions with perimetric wear: showing indentation marks, cracking (dotted arrow) and presence of encrusted Si-rich particles (arrows); (c) Used lower roll: detail of the oxidised region showing fainted surface finishing marks and silicon particle encrustation; (d) Used lower roll – detail of the regions with perimetric wear: showing indentation marks, cracking and presence of encrusted Si-rich particles. SEM-BEI.

Residues

Fig. 4a–d show the residues, which were collected on the roll/belt interface, indicating the presence of metallic wires along with pieces of eucalypt fibers (see Fig. 4a). The pasty lubricant residue (see Fig. 4b) was prepared for characterisation by placing the material on a sheet of paper, which was followed by residue homogenisation using a spatula; hot air drying; and stereomicroscope observation. The latter revealed the presence of metallic particles, being the critical regions sampled and gold metallised for further investigation in a scanning electron microscope. The helicoidal wires in the residue showed the presence of superficial deformation (see Fig. 4c), while the pasty lubricant residue revealed the presence of 15 μm diameter metallic particles (see Fig. 4d). EDS microanalysis (see Table 4) indicated that the metallic particles possess the same Cr/Ni ratio observed for the stainless steel belt. Additionally, the (stainless steel) helicoidal wires showed a higher Cr/Ni ratio, indicating its exogenous origin. X-ray diffractometry of the lubricant residue was carried out (*Shimadzu* diffractometer; Cu-K , radiation; accelerating voltage of 40 V; current of 40 mA; continuous scan; scan velocity of 0.5°/min; step of 0.02°; time of 2.40 s; and scan range between 40° and 140°) and the results, shown in Table 5, confirmed the presence of austenitic stainless steel particles and additional presence of silicon carbide particles – very likely to be the silicon-rich encrusted particles previously observed.

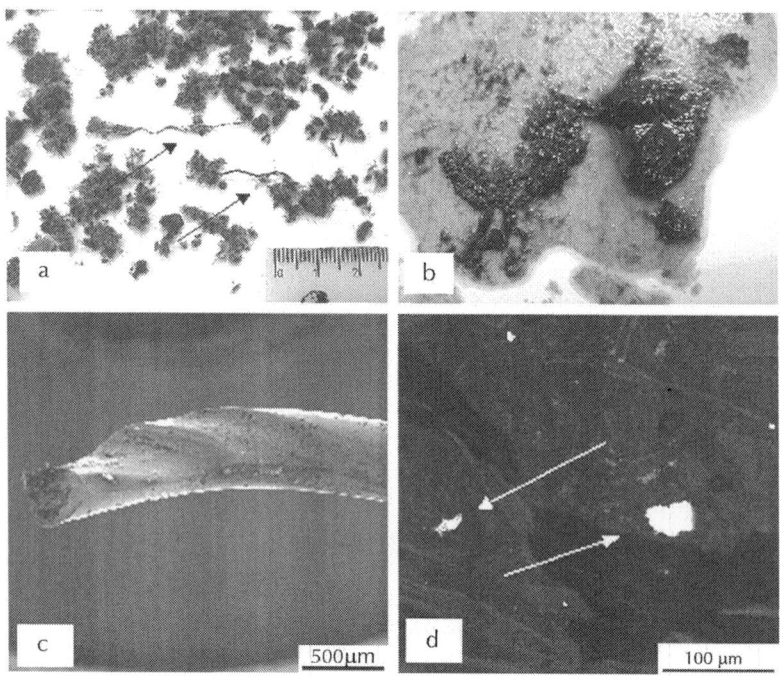

Figure 4: Residues. (a) Dry residue – sample 4 – featuring presence of helicoidal wires (arrows); (b) Pasty residue; (c) Detail of the helicoidal wire found in sample 4; (d) Detail of the metallic particles (arrows) found in samples 5 and 6.

Table 4: Chemical composition of the metallic material found in the residues – EDS microanalysis

Sample	Cr/Ni ratio	Result (wt. %)	
Wire	2.25	Fe	74
		Cr	18
		Ni	8
Metallic particles	1.86	Fe	75
		O	4
		Cr	13
		Ni	7

Stainless steel belt	1.93	–	–

Table 5: X-ray diffractometry results – lubricant residue

2ϑ	d (Å)	I (relative)	Phases[a]
77.4323	1.23157	100	Fe-γ(2 2 0), SiC(3 1 1)
43.9448	2.05875	81	Fe-γ(2 2 0) SiC(2 0 0)
64.3169	1.44722	64	SiC(2 2 0)
115.6853	0.90989	51	SiC
98.2411	1.01880	39	SiC(4 0 0)
136.3956	0.82964	30	SiC
111.1494	0.93385	23	SiC

[a]*Source*: JCPDS 2000 – Fe-γ-(Cr0.19Fe0.7Ni0.11) index card no. 33-0397; SiC – index card no. 49-1623.

Belts

Fig. 5a and b show the presence of a transversal through-thickness cracking, which features higher crack aperture and crack length values on the belt/roll interface, indicating that the cracking was nucleated on this interface. Fig. 6 and Fig. 7 show the microtopographic features of the belt working interfaces (belt/roll and belt/wood). The belt/wood interface (see Fig. 6a and b) shows scratches along the direction of movement and numerous parallel cracks, while the belt/roll interface (see Fig. 7a and b) shows the formation of large detachment pits. Microtopographic examination of the belt working interfaces confirmed that the mechanisms of superficial mechanical degradation are much more intense on the belt/roll interface; and that there is a relationship between cracking and the presence of pits. Fig. 8a–c show the results of macrostructural characterisation of the cracked belt, revealing the presence of parallel secondary cracks, while featured the same propagation path as the main crack – positioned at 45° from both working surfaces. The cracks formed on the belt/roll interface are, however, much deeper, confirming that the belt/roll interface is the critical one for the nucleation of transversal through-thickness cracking (500 µm against 10 µm). The macrostructural crack propagation from the belt/roll interface features the following path:

- 45° cracking until 500 µm deep ⇒ plane cracking between 500 µm and 750 µm deep ⇒ 45° cracking between 750 µm and 1.0 mm deep ⇒ plane cracking between 1.0 and 1.3 mm deep ⇒ 45° cracking between 1.3 and 2.2 mm deep ⇒ 30° cracking between 2.1 and 2. 4 mm deep ⇒ 45° cracking between 2.4 and 2.8 mm deep ⇒ 60° cracking between 2.8 and 3.0 mm deep.

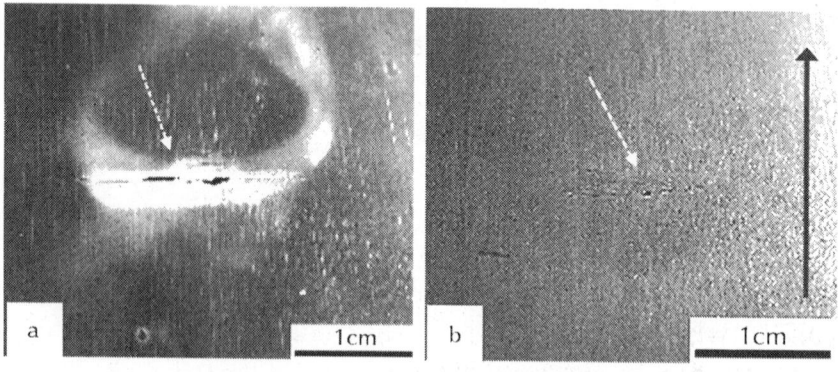

Figure 5: Used stainless steel belt – visual inspection. (a) Belt/roll interface: transversal through-thickness crack; (b) Belt/wood interface: transversal through-thickness crack (see dotted arrows). Large black arrow indicates the movement direction.

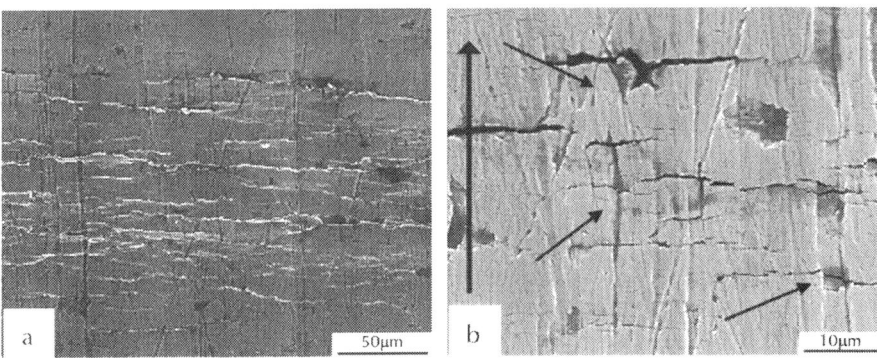

Figure 6: Used stainless steel belt – microtopographic examination – belt/wood interface. (a) and (b) Presence of numerous parallel secondary cracks perpendicular to the belt movement direction (see large black arrow). The cracks are parallel to the main crack and associated to the presence of indentation and scratching marks (see arrows). Observe presence of detached material leading to the formation of cavities (pits or spalls).

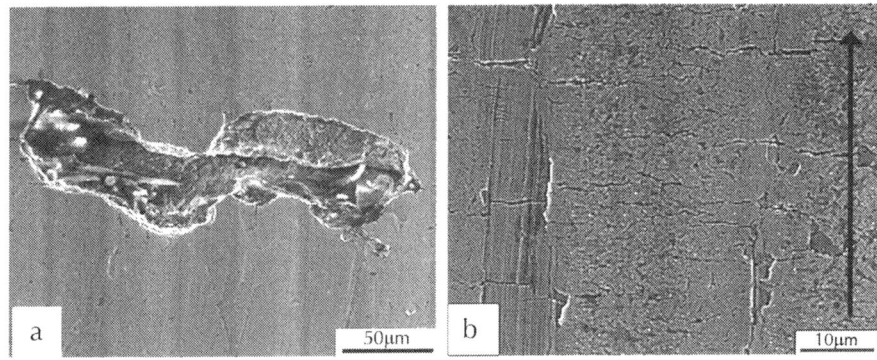

Figure 7: Used stainless steel belt – microtopographic examination – belt/roll interface. (a) and (b) Less intense presence of secondary cracks. This working interface shows, however, the formation of much larger pits or spalls.

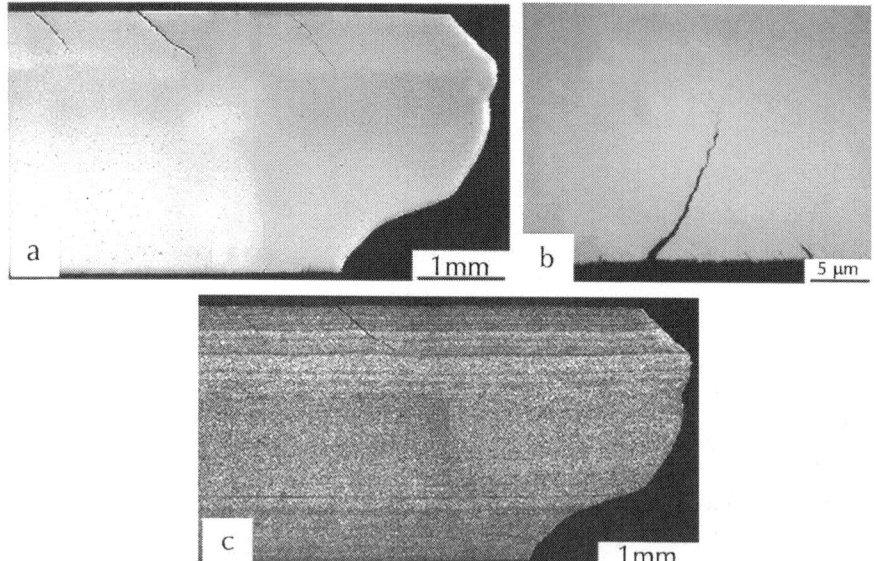

Figure 8: Used stainless steel belt – macrostructural examination. (a) Presence of 500 µm deep 45° secondary cracks at the belt/roll interface; (b) 10 µm deep 45° secondary crack at the belt/wood interface; (c) Etched macrostructure showing the formation of a wave-like deformation lines near the roll/belt interface.

Microstructural Characterisation

Roll and Residues

Fig. 9a–c show the microstructure of the roll, featuring the presence of the hardened martensitic case and a ferritic–pearlitic core. The working surface of the used rolls reveal the presence of a thin layer of deformed martensite (compare Fig. 10a and c); a 0.5 µm deep sub-superficial cracking associated with a thin layer of deformed martensite (see Fig. 10d); and a layer of iron oxide (37 at.%Fe–61 at.%O–1 at.%Si) – see Fig. 10b. Hardness profile measurements were carried out and the results indicated an effective hardened case depth (using reference value of 50 HRC) between 1.3 and 1.6 mm for the rolls. The new roll presented a higher value for the superficial hardness (~650 HV) against ~600 HV for the used upper roll and ~570 HV for the used lower roll. The required values of superficial hardness and effective hardened case depth were not made available for the present investigation. The microstructure of the helicoidal stainless steel wire found in the residue features a hardness between 600 and 630 HV and is formed by elongated austenitic grains (highly deformed) and non-metallic inclusions of aluminium oxide (fine series; globular oxide; level 3) – see Fig. 11.

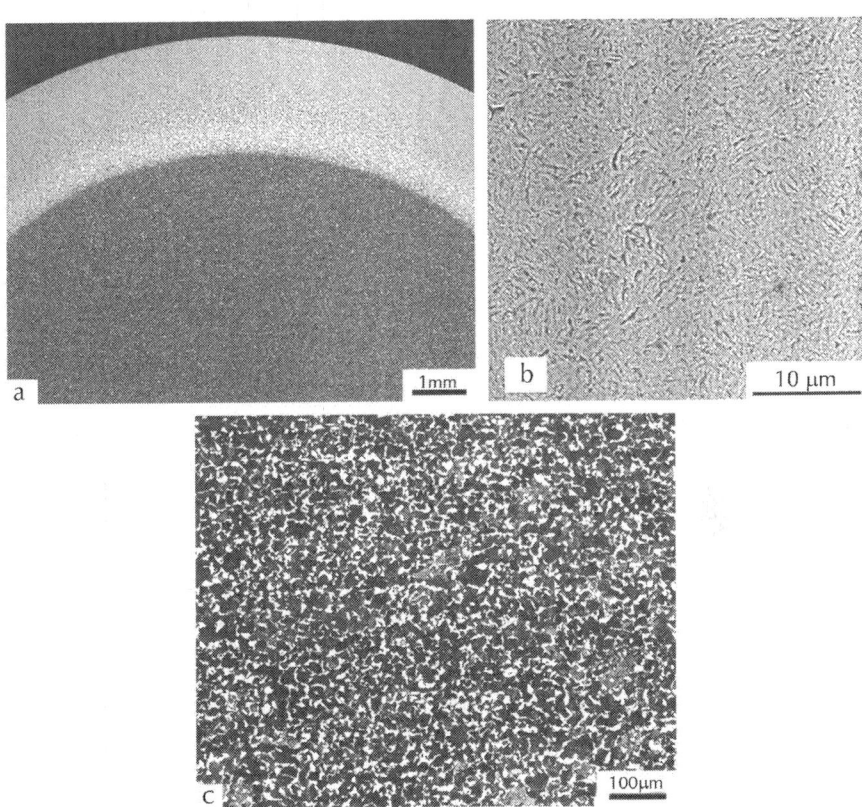

Figure 9: Roll – metallographic examination – etching: Nital. (a) Macro-structure showing the presence of a hardened case. (b) Hardened case: martensitic microstructure; (c) Core: ferritic–pearlitic microstructure.

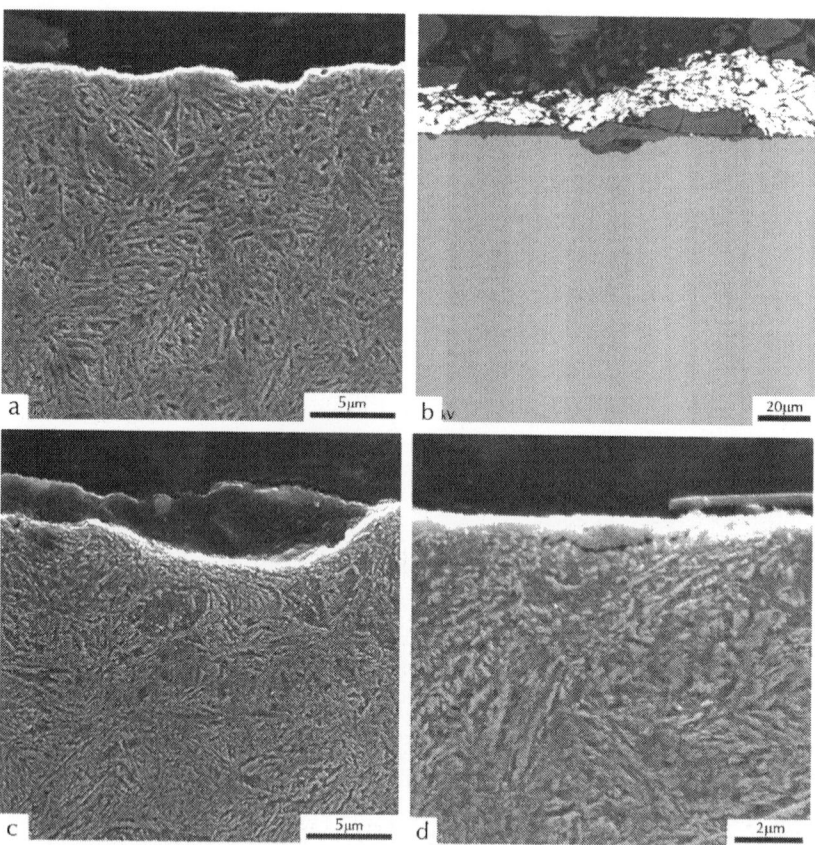

Figure 10: Roll working surface – metallographic examination – etching: Nital. (a) New roll showing martensitic microstructure; (b) Used roll showing superficial oxidation – iron oxide (grey); (c) Used roll showing deformed martensitic microstructure; (d) Used roll showing deformed martensitic microstructure with sub-superficial cracking. SEM-SEI and BEI.

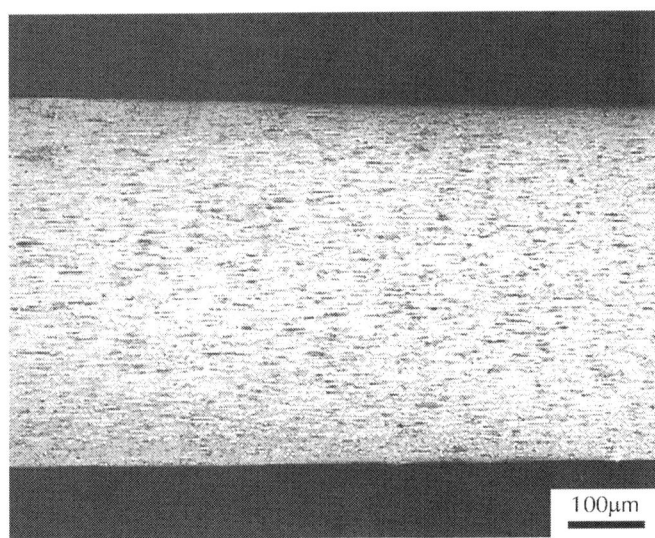

100µm

Figure 11: Helicoidal wire. Metallographic examination – Neophot light microscope. Etching: oxalic acid (electrolytic). General view of the deformed austenitic microstructure and presence on inclusions.

Belt

X-ray diffraction (tension: 40 kV A, current: 30 mA, Cr-Kα radiation; continuous scan – $\Delta 2\vartheta$ – between 30° and 160°, step: 0.02°, velocity: 1°/min) was carried out on a sample extracted from the cracked belt and the results indicated the presence of (1 1 0), (2 0 0) and (2 1 1) ferrite peaks and an unidentified peak (λ = 3.02644 A). There were no evidences of the presence of δ ferrite (tetragonal), martensite (tetragonal) nor γ austenite (FCC). The belt microstructure is formed by laths of ferrite and the presence of intergranular and interlath carbides (see Fig. 12a and b). A stripe of macrostructural segregation featuring a more refined microstructure associated to the presence of inclusions was also observed. Non-metallic inclusion content of the belt indicated the presence of D-type inclusion (globular oxide), thin series, and level 2. Finally, the presence of titanium carbo-nitrite particles was also observed. Microhardness Vickers testing (load of 300 and 500 g) was carried out from the belt/roll interface along the belt's thickness and the results are shown inFig. 13. The used belt generally features higher values of microhardness and the belt/roll interface shows the highest hardness increment, indicating that this interface is subjected to a more intense work hardening.

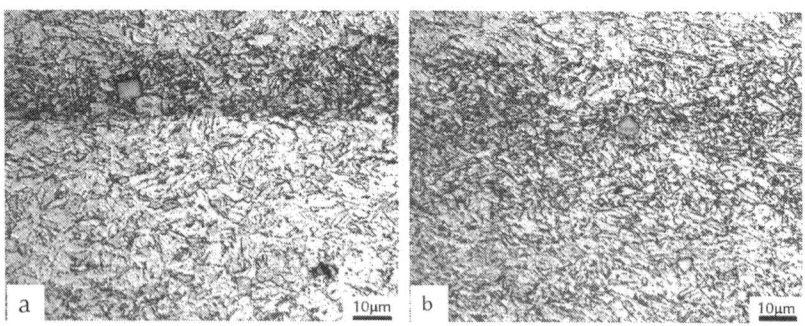

Figure 12: Belt – microstructural examination – used belt (a) and new belt (b) etching – modified villela. Ferrite with intergranular and interlath carbides. The plate macrostructure features banding, showing a more refined microstructure associated with the presence of inclusions.

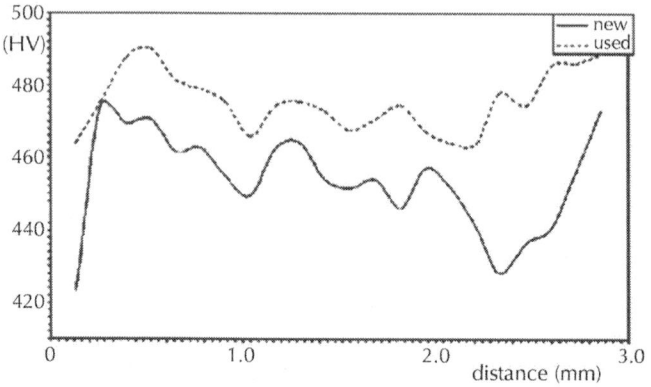

Figure 13: Microhardness profile results along the belt's thickness (starting on the belt/roll interface), comparing used (dotted line) and new (bold line) belts.

Microfractographic Examination – Belt

The microfractographic results of the transversal through-thickness cracking (after the exposure of the cracking surfaces by a cutting procedure) are shown in Fig. 14a–f. The fracture initiation region (see Fig. 14a and b) shows plastic deformation, without topographic hints for the identification of the crack propagation mechanism. Table 6 shows the EDS microanalysis results on the exposed crack surface, indicating that the crack initiation region (area 1) is more oxidised and richer in sulphur. These results suggest that the crack initiation region is a spall (width of 7.0 mm versus depth of 0.6 mm), which was formed by material detachment, followed by crushing and formation of iron oxide debris. Region 2 shows the presence of numerous ledges and parallel striation marks (see Fig. 14c–f), indicating that the crack propagation occurred by a fatigue mechanism nucleated at the spall cavity. The striation spacing increases from region 2–4 (see Fig. 14c–e). Finally, next to the belt/wood interface, the results indicated the presence of small conchoidal regions (region 5), which are formed by fine striation marks (see Fig. 14f), indicating that the parallel cracking observed on this interface is caused by a fatigue mechanism.

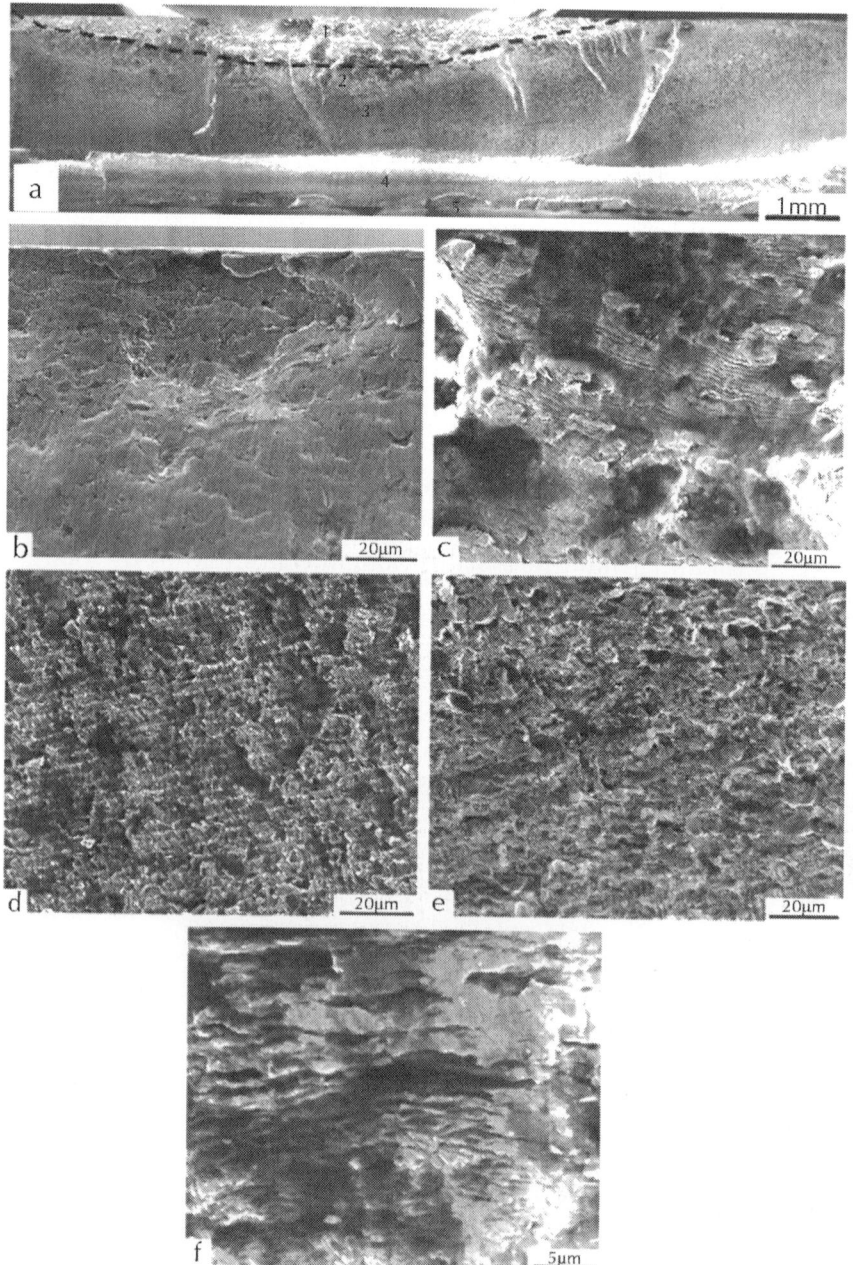

Figure 14: Fractographic examination-exposed surface of the transversal through-thickness cracking – SEM-SEI. Top: belt/roll interface; bottom:

belt/wood interface. (a) General view showing: region 1: formation of spall cavity (width = 7 mm and depth = 0.6 mm) and region 2: multiple crack nucleation (see formation of multiple nucleation ledges); (b) detail of the region 1 – spall surface – showing mechanical deformation; (c) detail of region 2 showing fine and parallel fatigue striation marks; (d) detail of region 3 showing parallel fatigue striation marks; (e) detail of region 4 showing broad and parallel fatigue striation marks; (f) detail of region 5 showing fine and parallel fatigue striation marks.

Table 6: EDS (energy dispersive spectrometry) [a] microanalysis results on the crack fracture surface

Elements	Belt/roll interface	Belt central region	Belt/wood interface
Oxygen (O)	31 ± 5	12 ± 1	9 ± 2
Silicon (Si)	1.9 ± 0.2	1.8 ± 0.2	2.1 ± 0.8
Sulphur (S)	1.7 ± 0.3	0.5 ± 0.1	0.4 ± 0.2
Chromium (Cr)	10.8 ± 1.4	12.0 ± 0.2	12.0 ± 0.6
Iron (Fe)	50 ± 4	66 ± 1	68 ± 2
Nickel (Ni)	3.2 ± 0.2	5.0 ± 0.1	5.0 ± 0.3

[a]Standardless, *area mode*, 100 s, tension: 15 kV, average of three analysis.

DISCUSSION

Visual inspection revealed the presence of a transversal through-thickness cracking, which was nucleated on the belt/roll interface. Microtopographic examination of the belt working interfaces confirmed that the mechanism of superficial mechanical degradation is more intense on the belt/roll interface of the belt and that there is a clear relationship between cracking and presence of superficial irregularities such as scratches, material detachment and indentation marks. Used belt also showed the presence of secondary cracks featuring a 45° crack-propagation path on both working interfaces, but the cracks formed on the belt/roll interface were much deeper, confirming that the belt/roll interface is critical for the nucleation and growth of transversal through-thickness cracking. Macrostructural examination of the stainless steel belts revealed the presence of oriented fibers along the rolling direction, with the used belt showing a more intense fibering. The experimental results indicated that the transversal through-thickness was promoted by the formation of a 600 μm deep contact fatigue spall, which was nucleated on belt surface in contact with the rolls (see Fig. 8 and Fig. 15) by a mixed rolling/slip contact fatigue regime. The spall finally acted as a preferential nucleation site for the nucleation of conventional fatigue cracks.

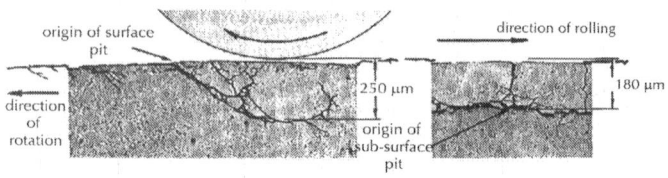

Figure 15: Spall formation by rolling contact fatigue (left – superficial nucleation; right – sub-superficial nucleation) [5].

In a generic way, the term contact fatigue refers to the processes of superficial damaged, which are caused by the repetitive superficial contact of two bodies, promoting formation of pits, wear debris and fatigue cracks [4]. In the case of a stationary roll/plate contact under normal loading (elastic indentation of a planar surface), where a circular cylinder of radius R (roll) is pressed against a planar surface (plate), the contact occurs over a strip of width $2a$ (Eq. (1)) along the cylinder axis. The maximum shear stress value (given by τ, see Eq. (2)) is located beneath the contact point, at depth (z) given by $0.78a$ (see Eq. (2)). Additionally, on the surface at the contact interfaces, the stresses are given by $\sigma_x = \sigma_y = -p(x)$ (see Eq. (3)). In Eq. (3) it is possible to see that the maximum stress is located at the centre of the contact, that is, at $x = 0$, and finally the maximum value is given by Eq. (4). By using Eqs. (1), (2), (3) and (4) it is possible to estimate values of shear and normal stresses, including the depth of the maximum shear stress in the case of a stationary roll/plate contact under normal loading.

$$a = \left(\frac{4PR}{\pi E^*}\right)^{\frac{1}{2}} \quad (1)$$

$$\tau_{\text{máx}} = 0.3 p_{\text{máx}} \quad \text{at } z = 0.78a \quad (2)$$

$$p(x) = \frac{2P}{\pi a^2} [a^2 - x^2]^{\frac{1}{2}} \quad (3)$$

$$p_{\text{máx}} = \left(\frac{PE^*}{\pi R}\right)^{\frac{1}{2}} \quad (4)$$

Where P is the applied normal load per unit length of the cylinder; $2a$ is the contact strip width (see Eq.(2) and Fig. 16); R is the cylinder radius; E^* is the equivalent *Young's modulus*; $p(x)$ is the normal stress distribution given by the *Hertz theory*; $x = z =$ depth.

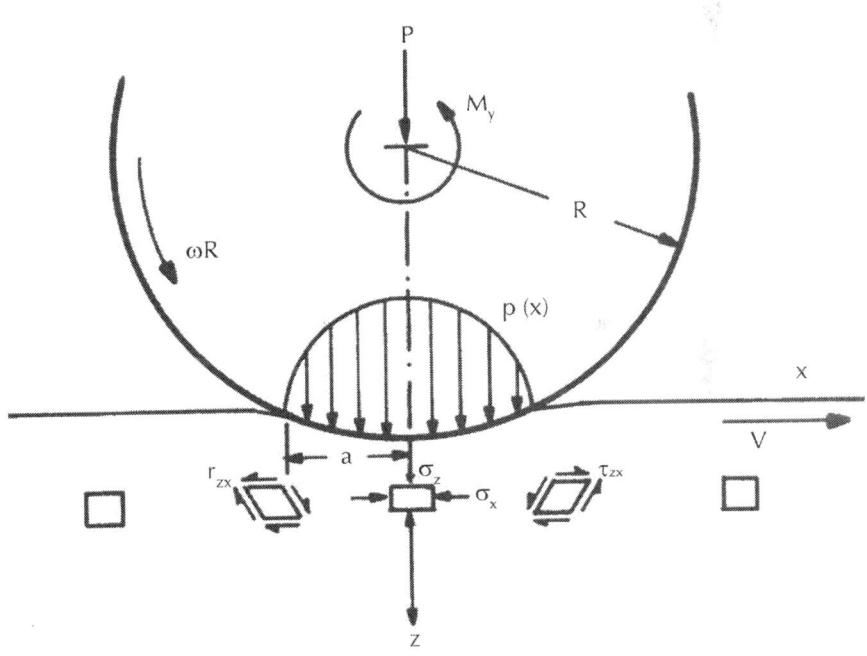

Figure 16: Rolling of a cylinder on a plane surface showing the stress fields. Note the reversal cycling of the shear stresses [11].

Assuming a work pressure of 4.7 MPa (or 470 N/cm²), which is applied over an area of 58 m² (20 m × 2.9 m) and supported by 1000 rollers (diameter = 18 mm and length = 2.9 m), the applied normal load per unit length of the cylinder (P) is equal to 94 kN/m, while p_{max} is equal to 619 MPa. The results indicated that the maximum shear stress ($_{max}$ equal to 185 MPa) should occur at a depth (z) equal to 75 μm. This result, however, was obtained considering that the load is perfectly distributed along all the roll extension, which is not very realistic as shown by the presence of perimetric marks shown on the rollers surface (Fig. 2b), indicating that the contact region is restricted to the region delimited by these marks (there are about 10 of these 5 mm wide perimetric marks in each roll and they are concentrated on the roll's extremities and centre). A new calculation of the depth of the maximum shear stress was made supposing that the contact load between belt and roll was not distributed over the entire length (2.9 m), but over 0.05 m (total length of the regions presenting perimetric marks) and the new results indicated that the maximum shear stress (σ_{max} equal to 1.4 GPa) occurs at a depth (z) equal to 573 μm. This extrapolation shows the importance of the contact area on the value and depth of maximum shear stress.

In addition, during the pure rolling of a cylinder on a plate (see Fig. 16 and Fig. 17) the elastic deformation from normal loading generates a contact area, whose features are described by the Hertz theory. Along the movement axis, it is observed just beneath the contact point that the shear stresses along x-, y- and z-axis are equal to zero and that the maximum shear stress is located at 45°. The material located just ahead of the rolling contact presents sub-superficially shear stresses along the axis (τ_{zy} et $_{yz}$). The same situation is observed for the material located just after the rolling contact, although the shear stresses, in this case, show an opposite direction, indicating the presence of a shear stress reversal cycling along the axis. The point of maximum τ_{zy} occurs at the same depth of τ_{max}, but the stress amplitude of this oscillation shows module equal to $2\tau_{zy}$, which is higher than τ_{max}. This cycling can promotes the formation of sub-superficial cracking, generally nucleated at inclusions or second-phase particles.

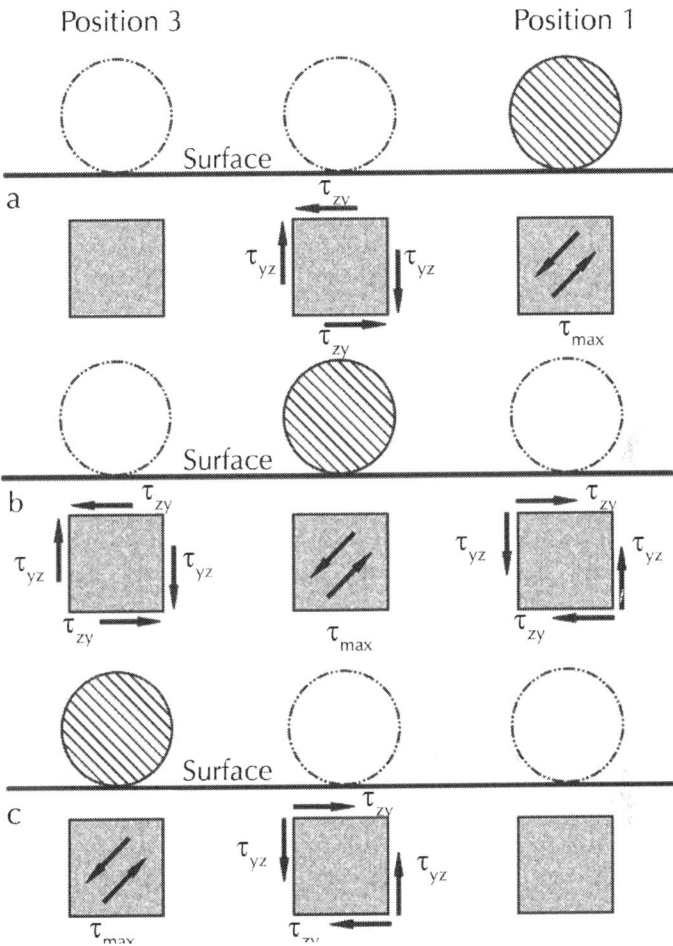

Figure 17: Rolling of a cylinder on a plane surface showing the stress fields. Note the reversal cycling of the shear stresses [6].

In the presence of frictional forces and sliding between the cylinder and the plate, tangential forces also changes significantly the magnitude and the stress distribution. In this case, there are tangential stresses on the contact surface in the opposite direction of the movement, as a consequence, the value of the maximum shear stress is increased and shifted towards the working surface [4], [5], [6] and [7]. In Fig. 3 and Fig. 6 it is very clear the presence of scratching on both belt and roll surfaces, indicating the occurrence of sliding between these surfaces. The experimental results also indicated that the contact fatigue cracks were nucleated on the surface of the belt in contact with the rolls and that the cracks propagated along a plane forming an angle of 45° to the surface (see Fig. 8). This angle agrees to the one calculated by Suresh [4] for gear teeth cracks nucleated by contact fatigue and found far from the pitch line. In contrast, cracks on the pitch line, where sliding goes to zero, the crack propagation angle is close to 20°. The angle , which is formed between the surface and crack propagation direction, indicates, therefore, the amount of sliding between two contacting bodies and, as consequence, the friction coefficient. An expression to the initial cracking angle is given by Eq. (5) [4].

$$\theta \approx \tan^{-1}\left[\left\{\frac{1-2v}{3\mu}+\frac{2(1+v)}{3}\right\}^{\frac{1}{2}}\right]$$

(5)

Where ϑ is the initial orientation of fatigue crack with respect to the free surface; v is the Poisson ratio; μ is the friction coefficient.

For a steel–steel contact and a friction coefficient varying from 0.3 to 1, Suresh [4] indicated that crack propagation angle should vary between 45° and 49°, which are close to ones observed during the experimental work. As a consequence, it seems reasonable to extrapolate that the value of the belt/roll tribo-surface friction coefficient is around 0.3. According to Williams [8], a friction coefficient value of 0.3 during the rolling of non-conformal surfaces is enough to shift the position of maximum shear stress to the tribo-surface, in the same way as fatigue wear during pure sliding [9]. Another factor that contributes for the increase of shear stress magnitude and its shifting towards the surface is the presence of dents on the working surface, particularly on the region of the pile up of the indentation. In this sense, the presence of exogenous or endogenous particles pressed between the contacting surfaces promotes superficial cracking nucleation [10]. In this sense, the experimental work indicated clearly that the used rolls presented perimetric wear marks and the presence of 2 µm silicon-rich encrusted particles (identified as silicon carbide), see Fig. 2 and Fig. 3. In addition, lubricant residues contained the presence of

- Helicoidal stainless steel wires, which were originated by the release of cleaning brush bristles;
- 15 µm diameter metallic particles, which were generated by material detachment of the belt.

All these foreign particles on the belt/roll tribo-interface contributed to an increase of the shear stresses on the surfaces and, consequently, to an increase of contact fatigue crack nucleation sites. These facts suggest that the lubrication system efficiency and the cleaning procedure should be optimised in order to increase life expectancy of the belt.

The experimental results indicated that the transversal through-thickness was promoted by the formation of a 600 µm deep contact fatigue spall, which was nucleated on belt surface in contact with the rolls (see Fig. 8) by a mixed rolling/slip contact fatigue regime. Finally, the fiber waving near the working interface observed in the used belt (see Fig. 8c) and the presence of striations marks (see Fig. 14c–f) on both surfaces of the belt's through-thickness cracking indicated the action of a cyclic bending loading. The spalling on the roll/belt surface acted as a stress raiser, promoting the nucleation and growth of "conventional" fatigue cracks, which led to the formation of a transversal through-thickness cracking.

CONCLUSIONS

- The transversal through-thickness cracking was nucleated on the belt/roll interface and that there is a clear relationship between the formation of cracking and the presence of superficial irregularities (scratches, material detachment and indentation marks) on working surface;

- Spalling was originated on the belt's surface (belt/roll interface) by a mixed rolling/slip contact fatigue mechanism;
- Conventional fatigue cracks, which led to the formation of transversal through-thickness cracks, were nucleated at about 600 μm deep spalling, indicating that contact fatigue played an important role on the nucleation of cyclic bending fatigue cracking;
- The contact rolls showed the presence of encrusted 2 μm silicon carbide particles, while the lubricant residues contained the presence of helicoidal wires and 15 μm diameter metallic particles, indicating that the system lubrication efficiency should be optimised in order to increase the life expectancy of the belt.

REFERENCES

1. Consultancy for carbon fixation. Wageningen University & Research, Netherlands. http://www.xs4all.nl/ ~ nlvdbeek/ report/ index.html.

2. Hiram Firmino. Eucalipto vilaõo ou heroí? Nem vilaõo nem heroí. Jornal do Brasil. On-line: http://jbonline.terra. com.br/jb/papel/   cadernos/jb_ecologico/2004/11/03/ jorjbe20041103006.html.

3. May Peter H. Forest certification in Brazil. In: Symposium forest certification in developing and transitioning societies: social, economic, and ecological effects. Yale School of Forestry and Environmental Studies. New Haven, Connecticut, USA. 10 & 11 June 2004.

4. Suresh S. Fatigue of materials, Chapter 3 – Contact fatigue: sliding, rolling and fretting. p. 435–82.

5. Metals handbook, vol. 10, Failure and prevention, ASM, 8th ed., Wear failure. p. 134 53.

6. Glaeser WA, Shaffer SJ. Contact fatigue. In: ASM handbook, vol. 19, Fatigue and fracture, 1996. p. 331–6.

7. Hyde RS. Contact fatigue of hardened steel. In: ASM handbook, vol. 19, Fatigue and fracture, 1996. p. 691–703.

8. Williams JA. Engineering tribology. Oxford University Press; 1994. p. 113.

9. Stachowiak GW, Batchelor AW. Engineering tribology. Elsevier; 2003. p. 572–3.

10. Girodin D, Ville F, Guers R, Dudragne G. Rolling contact fatigue tests to investigate surface initiated damage using surface dents. In: Beswick JM, editor. ASTM STP, 1419. West Conshohocken, PA: American Society for Testing and Materials; 2002.

11. Widner RL. Failure of rolling-element bearings. In: Metals handbook, vol. 11, Failure analysis and prevention, 1986. p. 490–513.

Chapter 3

The Manufacturing Process Parameters Affecting Color and Brightness of TiO² Pigment

Riyas Sharafudeen

Department of Chemical and Process Engineering Technology, Jubail Industrial College, Jubail Industrial City, 31961, Kingdom of Saudi Arabia

ABSTRACT

Background

Originally derived from colored minerals, inorganic pigment titanium dioxide is now highly engineered particles that imparts color or functionality to the objects in which they are used. Optical properties such as color, opacity, brightness, and gloss are important to users of paper and board grades. The color and brightness of TiO_2 sometimes get affected during its manufacture. These phenomena were observed in some commercial production plants. In order to investigate the effect of oil contamination on the pigment during its manufacturing process, such as from sand milling or larox filtration, the measurement on the brightness and color of titanium dioxide is made in this study. A correlation between the color and brightness values measured using different instruments is also included in this paper.

Results

It is observed that whenever there is an oil contamination in the pigment, the vehicle b values measured in acrylic lacquer paint film of the pigment is affected. BYK L* is poorly correlated with the HunterLab vehicle L, and the dry Hunter b is in correlation with the DBC BYK b*, whereas L has no correlation with DBC BYK L.

Conclusion

The contamination of pigment with lubricating oil from the machinery used for the manufacturing process influences both dry and vehicle b values. It is not possible to use HunterLab in place of BYK for the dry L and b measurements.

BACKGROUND

Titanium dioxide is the most widely used white pigment because of its brightness and very high refractive index, in which it is surpassed only by a few other materials. TiO_2 is unique because it efficiently scatters visible light, thereby imparting whiteness, brightness, and opacity when incorporated into a coating. Dry compacted TiO_2 samples are characterized by their brightness and whiteness and exhibit reflectance properties approaching that of the perfect reflecting diffuser. 'Color' is carefully controlled during the TiO_2 manufacturing process through the removal of the trace amounts of metal oxide contaminants. These light-absorbing contaminants will detract from the brightness and whiteness of the pigment and can affect the appearance of white and near-white paints containing that pigment. The opacity of white pigments mainly derives from the scattering, while the opacity of black pigments is a result of absorption ability [1-5]. It was observed that if the TiO_2 is produced in a commercial production plant on particular days when there is oil leak in the larox filter, the color and brightness of the produced pigment failed when measured the same in acrylic lacquer paint (vehicle b). Also, this failure is not seen when the color and brightness are measured in dry mode (dry b). These observations were not a single case but were seen occasionally in the quality analysis. The failure in the vehicle color, once commenced, was seen continued consecutively with three to four final products. The sources of oil contamination during the manufacturing process can be as follows: In the chloride process of manufacturing TiO_2, rutile or high-grade ilmenite is converted to titanium tetrachloride ($TiCl_4$) gas. The conversion takes place in a chlorinator (i.e., fluidized bed reactor) in the presence of chlorine gas at 850°C to 950°C, with petroleum coke added as a reductant. The chief reaction products are volatile metal chlorides, including $TiCl_4$, which are collected. The nonvolatile chlorides and the unreacted solids that remain in the chlorinator are wasted, forming the special waste stream 'chloride process waste solids'. The gaseous raw product stream is purified to separate the titanium tetrachloride from other chlorides. Separation is by fractional condensation, double distillation, and chemical treatment. Ferric chloride ($FeCl_3$) is removed as a major acidic liquid waste stream through fractional condensation. Additional trace metal chlorides are removed through double distillation. Finally, vanadium oxychloride ($VOCl_3$), which has a boiling point close to that of $TiCl_4$ (136°C), is removed as a low-volume non-special waste by complexing with mineral oil and reduced with hydrogen sulfide to $VOCl_2$ or by complexing with copper. The purified $TiCl_4$ is then oxidized to TiO_2 at 985°C, driving off chlorine gas, which is recycled to the chlorinator. Aluminum chloride is added in the oxidation step to promote formation of the rutile crystal which is the TiO_2 product. The next stage of TiO_2 manufacturing is surface treatment, where the TiO_2 from oxidizer is converted to slurry in water, which is then subjected to sand milling to the required particle size. The clarified slurry from the sand mill is treated with alumina or zirconia for coating the surface of bare titanium dioxide with amorphous aluminum or zirconium oxide. The surface-treated product is then filtered through larox by pressing and followed by organic treatment and spin flash drying. Finally, the dried product is subjected to micronization process to get the surface-treated product. The brightness and color of this product is important for its marketing.

The possible sources of oil contamination are during sand milling and larox filtration process. The oil leak is due to lack of monitoring system and improper maintenance. The oil used in the sand mill rotator for avoiding friction can leak due to its continuous use. Likewise, the larox oil can leak during the pressing process, which can contaminate the filtered cake. In order to investigate this issue, it was decided to study the impact of oil contamination on the pigment, color, and brightness values and correlate the color and brightness values measured using different methods.

This paper explains how the pigment L and b values measured in dry and vehicle are influenced by oil contamination during the manufacturing process and determine a correlation between the HunterLab (Reston, VA, USA) L and b with the BYK DBC L and b in order to identify the pigments that failed in vehicle b using HunterLab with the dry b* measured using BYK instrument (BYK-Gardner, Columbia, USA). The correlation between the dry HunterLab values with the dry BYK is also included in this study; the TiO_2 samples which failed in acrylic lacquer paint b values were collected and measured in terms of brightness and color by both BYK and HunterLab instrument as per the above procedure for this correlation study.

METHODS

Materials

Chemicals used in the study includes the raw slurry of TiO_2 from the commercial plant, Larox oil RENOLIN B (Fusch Petrolube, Shah Alam, Selangor Darul Ehsan, Malaysia), grade 46, acrylic resin A21 (Rohm and Hass, The Dow Chemical Company, Bristol, PA, USA), Santicizer 160 (Monsanto Chemical Company, St. Louis, MO, USA), commercial grade toluene, commercial grade methyl ethyl ketone, methyl n-amyl ketone, standard TiO_2, Ektasolve EB acetate (Eastman Chemical Company, Kingsport, TN, USA).

Apparatus and Equipments

The experiments were carried out using Wig-L-Bug grinder with timer (Sigma-Aldrich Corporation, St. Louis, MO, USA), 50 ml polyethylene Nalgene vials (Nalge Nunc International Corporation, Rochester, NY, USA), glass beads, white cards, bird blade film applicator, 6 mm and 6 in. (15.2 cm) path, vacuum plate, Mylar film, cut from roll of approximately 2–3 in.² area, metal cylinder, metal plunger, metal base with a smooth top, hydraulic press, HunterLab and BYK colorimeter and laboratory larox (Labox).

Hunterlab Color Measurement in Acrylic Lacquer Paint

Paint mixtures and the standard TiO_2 were prepared for each sample by weighing the following ingredients into a 50-ml polyethylene vial: 20.0-g size (203) glass beads, 5.0 g of TiO_2 sample, and 26.9 g of master batch prepared in acrylic lacquer-13.2% PVC and grinded for 15 min on the Wig-L-Bug. The beads were allowed to settle for 5 min, and one white card was placed on the vacuum plate. A side-by-side drawdown of the sample and standard was made. The second card was placed on the vacuum plate and another drawdown was made, reversing the order of the sample and the standard. The standard and the sample on each card were labeled. There are no void streaks and each film is wide enough to fill the color meter light port, and a sufficient amount of paint has been used to make the drawdown to produce the desired film thickness. The samples were air-dried for 1 h on a flat, draft-free, dust-free surface and were measured for brightness and color using standardized HunterLab color difference meter for L and b scales. The average the L and b values for each standard/sample pair was calculated as

Average value = average sample reading − average standard reading + standard value.

Color Measurement of Dry Compressed TiO$_2$

An organization called Commission Internationale de l'Eclairage (CIE) determined the standard values that are used worldwide to measure color. The values used by CIE are called L*, a*, and b*, and the color measurement method is called CIELAB.

Symbol L* represents the difference between light (where L* = 100) and dark (where L* = 0). A* represents the difference between green (−a*) and red (+a*), and b* represents the difference between yellow (+b*) and blue (−b*). Using this system any color that corresponds to a place on the CIELAB coordinate system was shown in Figure 1. The variables of L*, a*, b*, or E* are represented as delta L*, delta a*, delta b*, or delta E*, where delta E* = delta (delta L*^{2} + delta a*^{2} + delta b*^{2}). It represents the magnitude of the difference in color, but does not indicate the direction of the color difference.

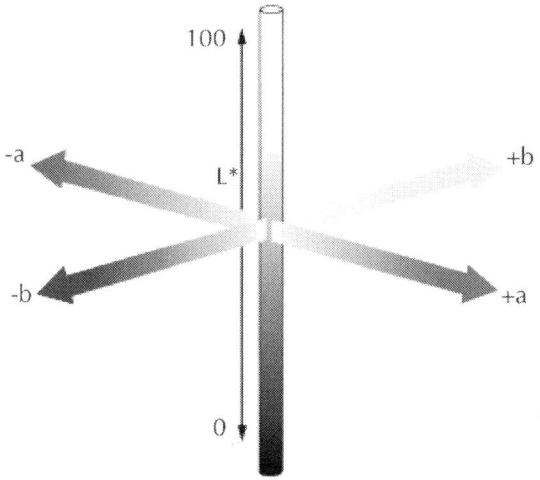

The CIELAB color coordinate system

Figure 1: The CIELAB color coordinate system.

A 2–3 in.² of Mylar film was placed on top of a smooth metal base, and a metal cylinder was filled about full with the TiO_2 sample to be tested, covered with a metal seal, and applied 5,000 psi for 30 s. The pressure was released, the cylinder was removed, and the face of the pellet was inspected for smoothness and measured for dry color and brightness using HunterLab. The standard TiO_2 sample was also prepared by the above procedure. For comparison, the dry L and b values were measured using BYK color-view meter as well.

RESULTS AND DISCUSSION

An important observation during the filtration was that the oil was coming out from the system along with the filtrate. The filtrate collected appeared to be a suspension of oil in water. Even though the majority of the oil was visually seen passing with the filtrate, yet a trace amount of oil was found adsorbed on the pigment surface, giving a distinct odor to the pigment. The dry and vehicle L and b values of the samples were tested as mentioned above and the results are tabulated in Table 1.

Table 1: Dry and vehicle L and b values of the samples

Experiment No	Sample	Dry				Vehicle	
		BYK		HunterLab		HunterLab	
		L	b	L	b	L	b
1	Reference sample	98.33	1.63	97.24	1.29	95.40	0.68
	Oil-contami-nated sample	98.37	1.73	97.27	1.41	95.45	0.76
2	Reference sample	98.18	1.60	97.34	1.27	95.20	0.66
	Oil-contami-nated sample	98.24	1.71	97.49	1.43	95.22	0.77

Sharafudeen

Sharafudeen International Journal of Industrial Chemistry 2012 3:26, doi: 10.1186/2228-5547-3-26

Hunterlab Results

It is clear from the above results that there is no considerable change in the dry L value of the pigment upon oil contamination, while the dry b values increased by 0.12 and 0.16 units, respectively, in oil-contaminated samples. Still, the values are within the specification limits for the final marketing product and therefore, the oil-contaminated pigment passed the dry b test using HunterLab. However, when the same samples were subjected to the vehicle b test, an increase of 0.08 and 0.11 units was seen. This increase, though less compared to the increase in dry b value, leads to the failure of the pigment in the vehicle b test. Since the specification for vehicle b value is very narrow, the weight of smaller difference in vehicle b will be really higher than that of dry b values. Also, the gloss of the paint film measured was found to be more than expected. This may be due to the increase in specular reflection due to the presence of oil, and the diffused reflections may be reduced, which appear as a low-brightness value. The specular and diffuse reflection is shown in Figure 2. The brightness of a pigment is due to diffuse reflection while the gloss is due to specular reflection.

Specular Reflection   Diffuse Reflection

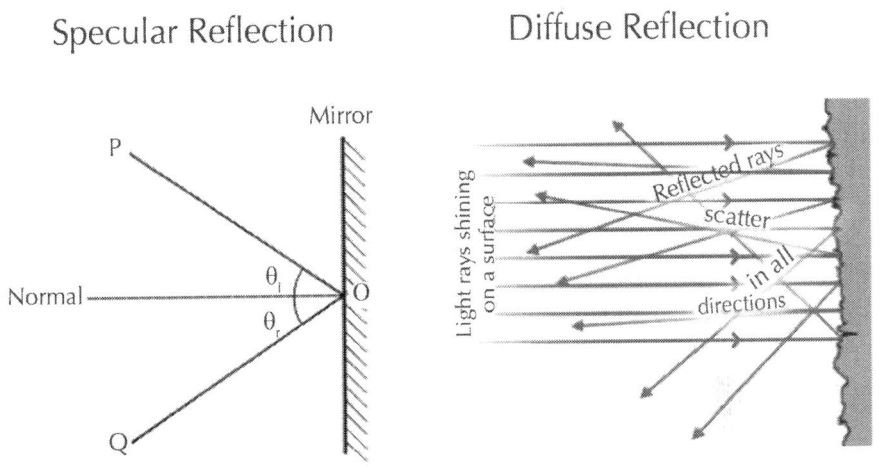

Figure 2: The specular and diffuse reflection.

Effect of Temperature on Oil-contaminated Samples

When the air-micronized samples were subjected to heating in air oven at 100°C and 180°C for 1 h, the oil-contaminated pigment was found to change in color towards yellow. This effect was more when the temperature was raised from 100°C to 180°C. Both dry and vehicle L and b values were measured by HunterLab and BYK. The results of dry and vehicle L and b values of the laboratory-prepared samples that are subjected to heating in air oven at different temperatures are given in Table 2.

Table 2: Values of the laboratory-prepared samples that are subjected to heating in air oven at different temperatures

Sample	Dry				Vehicle	
	BYK		HunterLab		HunterLab	
	L	b	L	b	L	b
Reference sample heated at100°C/1 h	98.11	1.63	97.16	1.30	95.08	0.77
Reference sample heated at 180°C/1 h	98.14	1.67	97.11	1.71	94.96	0.99
Oil-contaminated sample heated at 100°C/1 h	98.15	1.74	97.46	1.41	95.11	0.83
Oil-contaminated sample heated at 180°C/1 h	98.14	2.02	97.35	1.84	95.06	1.08

Sharafudeen

Sharafudeen International Journal of Industrial Chemistry 2012 3:26, doi: 10.1186/2228-5547-3-26

In all the samples, the L values remained more or less the same upon heating; while both dry and vehicle b values were increased noticeably upon heating. It is also observed that the increase in dry b value is more pronounced in HunterLab than BYK. However, the increase in dry as well as vehicle b values upon heating is more or less the same in the reference sample and the sample contaminated by oil. So it is not possible to distinguish between reference and oil-contaminated samples from this result. In other words, both reference and oil-contaminated samples heated at 100°C passed the dry b test, while when the temperature was increased to 180°C, both samples failed the dry b test. Unlike this observation, all the four samples failed in the vehicle b test irrespective of the nature of the sample. So, it is obvious from this result that the heating did not make any special change in the sample contaminated by oil; rather, it changed the b value of both reference and oil-contaminated sample to the same extent. The brightness and color of oil-contaminated pigment were compared with the pure, uncontaminated pigment prepared from the same base pigment under similar conditions. One of the reasons for this was expected to be the oil contamination during the filtration stage due to the possible leakage of lubricant oil from the larox filter pressing stage; hence, the same larox lubricating oil, RENOLIN B, grade 46 was doped with the pigment during filtration. Another source for oil contamination is from the sand mills during the grinding process. The gear oil used in such mills can leak and contaminate the slurry which proceeds to the surface treatment process. In order to avoid this oil contamination, an effective monitoring of all the sand mills and larox filter for oil leak is required.

For comparison, all these samples were subjected to dry L and b tests using the new BYK color-view meter. Similar to HunterLab results, there was no noticeable change in dry L value; while the dry b value was increased in oil-contaminated samples by 0.1 and 0.11, respectively. Unlike HunterLab dry b values, the b values measured by BYK were above the present specification limit and, therefore, the oil contaminated samples failed the dry b test by BYK. So correlations between the two instruments readings are necessary.

Correlation between Color and Brightness Measured with Hunter and BYK Colorimeter

Correlation of Dry Color and Brightness Measured with BYK color-View with HunterLab

Correlation between Vehicle b with BYK b*

It is found from the study that almost all the TiO_2 samples failed in vehicle b showed dry b value of around 2 and above. The same trend was observed for lot samples also. Figure 3 represents the variation of vehicle b values with BYK b* for samples. It is clear from the graphs that for the bagging bin samples R^2 was found to be 0.583, hence, it can be inferred that there exist a good correlation between vehicle b and BYK b*for samples.

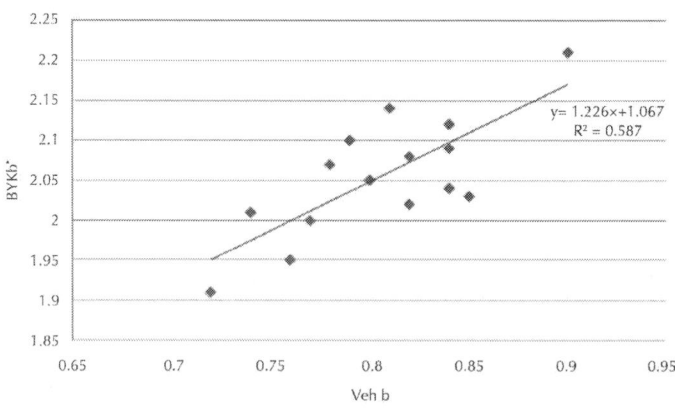

Figure 3: Variation of vehicle b values with BYK b* for samples (indicated by blue diamonds).

*Correlation between Vehicle L with BYK L**

As a variation of vehicle L with BYK L* for the TiO$_2$ samples in Figure 4, the vehicle L of Figure 3samples were plotted against the BYK L*. It is evident from the figure that the R^2 value of the bagging bin samples is 0.162 which indicates that the correlation between vehicle L and BYK L* is very poor. Hence, the correlation is poor in the case of L value.

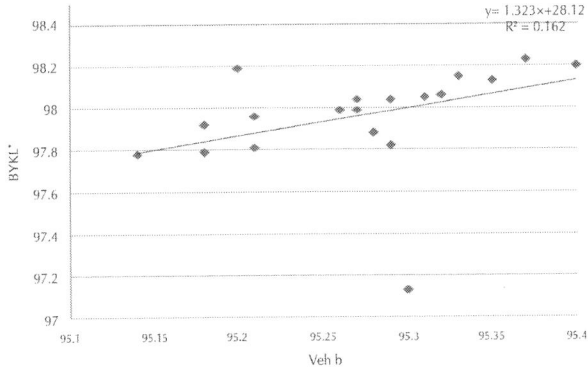

Figure 4: Variation of vehicle L with BYK L* for samples (indicated by blue diamonds).

Correlation of Dry Block Color BYK with Dry HunterLab L and b

Correlation between Dry HunterLab b with BYK b*

It is observed that the dry HunterLab b is in good correlation with BYK b*. As it is clear from the Figure 5, the variation of dry HunterLab b with BYK b* for the TiO_2 samples of R^2 was found to be 0.837 which can be inferred that there exist a good correlation between dry HunterLab b and BYK b*.

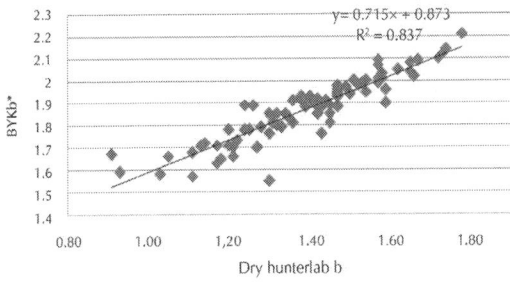

Figure 5: Variation of dry HunterLab b with BYK b* for samples (indicated by blue diamonds).

Correlation between Dry HunterLab L with BYK L*

Figure 6 represents the variation in dry hunter lab L with the BYK L*. It is evident that the R^2 is 0.382 which indicates that the correlation between them is poor.

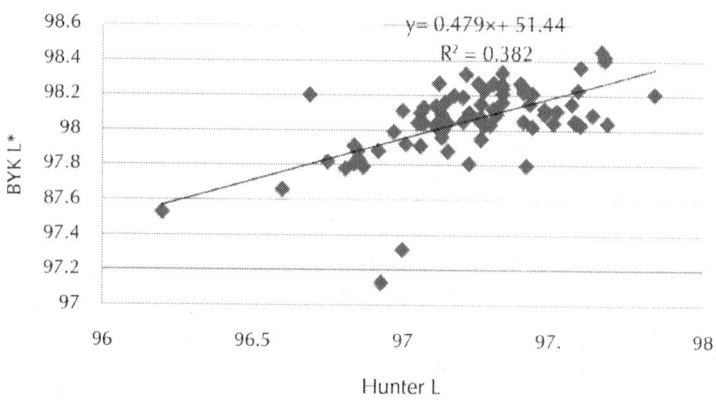

Figure 6: Variation in dry HunterLab L with the BYK L* (indicated by blue diamonds).

EXPERIMENTAL

Experimental

Sample used in this study was prepared by the following procedure. A 0.5 ml of oil was contaminated with RENOLIN B, grade 46 which is the same oil used in the hydraulic press in the plant larox filter. A 300 ml of surface-treated slurry of titanium dioxide was filtered using laboratory larox (Labox), was dried after filtration, and was micronized in the laboratory air micronizer then measured in its brightness and color. Another reference sample was prepared without oil contamination. Since the laboratory micronizer utilizes air while the plant micronizer utilizes super-heated steam, it was decided to heat the air-micronized samples at 100°C and 180°C for 1 h to see the effect of temperature on oil-contaminated samples. The experiment was repeated for confirmation. The brightness and color were measured using instruments HunterLab and BYK. The dry acrylic lacquers L and b were measured using HunterLab and BYK colorimeters [6,7].

CONCLUSIONS

The contamination of pigment with lubricating oil from the machinery used for the manufacturing process influences both dry and vehicle b values. The oil contamination can be identified by dry b test, if the HunterLab dry b specification limits are made narrow. Since part of this oil can be carried with the filtrate during filtration process, the effect can impact the succeeding batches of treatment, since the filtrate is recycled to the treatment tank. This effect would be continued until the oil is completely removed from the system. The dry b test by BYK is promising. It is clear that while setting the upper specification limit of BYK b*, it is good to consider in-process samples; if the upper specification limit for BYK b* is set at 2.0, then it is possible to identify samples which fail in the present vehicle b test due to any process contaminations.

From the correlation study, it can concluded that BYK b* is in good correlation with the HunterLab vehicle b values and when it exceeded 0.70, the corresponding BYK b* reaches 1.9 or higher, and from this it can be inferred that the sample which failed in vehicle b test can be identified with the BYK b* value. the BYK L* is poorly correlated with the HunterLab vehicle L, and the dry Hunter b is in correlation with the DBC BYK b*, whereas L has no correlation with DBC BYK L*; hence, it is not possible to use HunterLab in place of BYK for the dry L and b measurements.

ACKNOWLEDGMENTS

I express my sincere thanks to Mr. Hassan Mones El-Dekki, the general manager of the National Titanium Dioxide Company, Saudi Arabia for providing the necessary facilities and resources for this study, and to all members of the research and development department for their valuable help.

REFERENCES

1.   Kirk O (1997) Kirk-Othmer encyclopedia of chemical technology Wiley, New York p 4

2.   Lambourne R (1987) Paint and surface coatings Camelot Press, Southampton

3.   Billmeyer FW Jr, Salzman M (1981) Principles of color technology Wiley, New York

4.   Wyszecki G, Stiles WS (1982) Color science Wiley, New York p 2

5.   Patton TC (1973) Pigment handbook Wiley, New York

6.   HunterLab (2008) Application note. Test methods for color measurement. HunterLab 8:4

7.   Buxbaum G, Pfaff G (2005) Industrial inorganic pigments Wiley-VCH, Weinheim

Chapter 4

The Research on Failure Analysis of Fluid Cylinder and Fatigue Life Prediction

GuoRong Wang[a], LinYan Chen[a], Min Zhao[b],
Rong Li[c], and BenSheng Huang[d]

[a]School of Mechatronic Engineering, Southwest Petroleum University, ChengDu 610500, China

[b]Great Wall Engineering Research Institute, China National Petroleum Corporation, PanJin 124010, China

[c]SJ Petroleum Machinery Co., JiangHan Petroleum Administration Bureau of Sinopec, JingZhou 434024, China

[d]School of Materials Science and Engineering, Southwest Petroleum University, ChengDu 610500, China

ABSTRACT

To analyze the reasons of fluid cylinders' rupture, macro-analysis, SEM, composition inspection, metallographic analysis, hardness test and mechanics performance test of fluid cylinders materials were implemented. Two different kinds of fatigue life prediction methods have been proposed which are based on total life analysis and strain–life methodology. The results indicate that: the failure cylinders' material quality is satisfactory. Fatigue damage caused by high working, stress and corrosion is the main reason of cracking. The fatigue life prediction illustrates that strain–life methodology is well adapted to fluid cylinders.

INTRODUCTION

With the development of fracturing technology, the requisite high pressures and rates result in more frequent equipment maintenance and repair. The unexpected in-service failure cases of frac-pump's fluid cylinders motivate the present research. The complex structure and harsh working environment, for instances, high repeating stress, wear, erosion, and corrosion, lead to life short [1]. The serious failure of fluid cylinders increased the expenditure of oil and gas, influenced fracturing operation efficiency and even brought serious danger to operation, which have become a bottleneck in promoting the performance of frac-equipment.

Nowadays, the failure problem of static load components is rare and prediction or prevention measures are reliable on this. What is more, most mechanical components failure originates from fatigue load, so fatigue reliability analysis and prognosis are critical to structure design and maintenance [2]. But, the fatigue prediction is a tough process, which is closely related to geometrical structure, materials attribute, load type, service condition, etc. Moreover, fatigue failure often occurs without any warning signs. Still, extensive efforts have been made on the deterministic analysis of fatigue damage: Niesłony et al. [3]utilized Manson–Coffin–Basquin and Ramberg–Osgood laws for strain-based fatigue life assessments; Liu and Mahadevan [4] explained that fatigue crack extension is a random process in nature and various uncertainties, which required accurate fatigue reliability analysis and life prediction, the simulation-based methods include the direct Monte Carlo (MC) simulation method and the MC approach with different sampling techniques [5] and [6] is the efficacious means for fatigue prediction; actually, fatigue reliability is a time-dependent reliability problem and different metrics can be used to describe the random nature of fatigue damage [7]. One common approach is to calculate the failure probability at a specified life level. Zhao et al. [8] tested the effect of self-reinforcement by explosion and the experimental results showed that autofrettage can increase the fatigue life by around 5 times, but their testing model have been simplified as thick-walled cylinder; Tao et al. [9] used nonlinear analysis to identify the autofrettage pressure of fluid cylinders, however, when estimating the fatigue life, he only used empirical formula without considering load type and load frequency, so the fatigue life could not be increased for one hour, which lead to weak practicability in site. Besides, FEM technology has been widely used to simulate and analyze the fatigue life of fluid cylinders. Shi et al. [10], Zhang et al. [11] and He et al. [12] all hold that fatigue damage is easily occurred in the bore intersection area, but the failure reasons of fluid cylinders did not explain.

The main purpose of this paper is to figure out the failure reasons of cracking, crack propagation mechanism as well as to explore the approach of fatigue evaluation for fluid cylinders on the basis of the scrapped fluid cylinder with type of 5ZB-1860. And failure analysis experiments are carried out. Otherwise, three different kinds of fatigue life prediction methods have been introduced, which are based on total life analysis, the strain–life methodology, and a recently developed crack growth-based [13] fatigue life prediction model.

THE FAILURE MODE OF FLUID CYLINDERS

The fluid cylinders' failure modes are various. The cracking, which almost accounts for 80%, is the major and the most dangerous mode according to statistical analysis by field technicians.

It has been found that the oil field generally use straight-through fluid cylinders, namely, the axis of the discharge valve and the suction valve are in the same line through oilfield investigation. The fluid cylinders were produced by Halliburton, SJ petroleum machinery co., Baoji oilfield machinery co., etc. Though they were manufactured by different companies, the failure modes were similar. The cracks mostly appeared at six areas as shown in Fig. 1: mark '1' is inhaled cap, where often emerges circumferential crack, and this issue only fit for outer suction packing bore, as shown in Fig. 2(a); and mark six is suction area, bore intersections marked as '2–5', these locations often initialize crack source area and the cracks grow to long linear cracks along with axis direction discharge valve and plunger respectively as shown in Fig. 2.

Figure 1: The cut-open view of fluid cylinder.

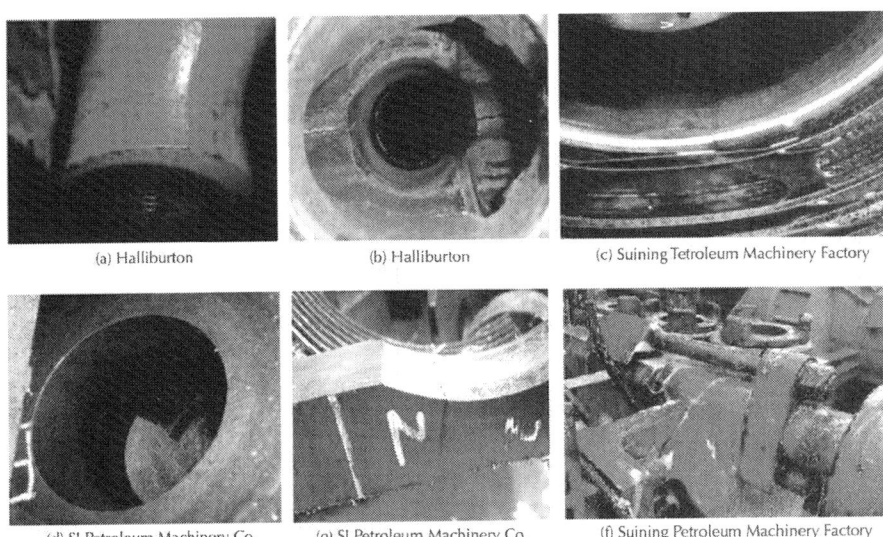

(a) Halliburton   (b) Halliburton   (c) Suining Tetroleum Machinery Factory

(d) SJ Petroleum Machinery Co.   (e) SJ Petroleum Machinery Co.   (f) Suining Petroleum Machinery Factory

Figure 2: Failure fluid cylinder.

Based on subordinate units Downhole Service Company of CNPC Chuanqing Drilling Engineering Co.Ltd, this paper collected some failed fluid cylinders' pictures and the working time of some types were listed in Table 1, from which it can be obtained that the pumps' service life is generally short, mostly less than 200 h.

Table 1: The working life for fluid cylinder

Number	Fluid cylinder replace time	Manufac-turer	Fluid cylinder type	Total working time of frac-pump (h)	Total work-ing time of fluid cylin-der (h)
1	10/31/2010	SuiNing Petroleum Machinery Factory	2000HP	1332.0	32
2	7/8/2010	Halliburton	HQ-2000	993.5	156.5
3	12/30/2010	Halliburton	HQ-2000	966	8.5

4	8/29/2010	SJ Petroleum Machinery Co.	5ZB-1860	1320.5	123
5	10/3/2010	Halliburton	HQ-2000	1407.7	56.67
6	12/24/2010	SuiNing Petroleum Machinery Factory	2000HP	1002	67
7	9/17/2010	Halliburton	HQ-2000	1022.4	73.42
8	10/27/2010	SuiNing Petroleum Machinery Factory	2000	1383.0	50
9	4/11/2010	SJ Petroleum Machinery Co.	5ZB-1860	1210.0	137.5
10	9/28/2010	SJ Petroleum Machinery Co.	5ZB-1860	1040	153

FAILURE ANALYSIS

Fracture Analysis

The macro-morphology of the failed fluid cylinder is shown in Fig. 3, which is observed by naked eye firstly. Besides two linear cracks have been found at bore intersection bore. Two cracks' length is about 50 mm. The two cracks began at the area of intersection bore and propagated to the discharge valve along with the axial direction of discharge valve. The extension direction of crack 'a' and crack 'b' with circumferential orientation of the fluid cylinder both have angles of 90°. Fracture type of the fluid cylinder cracks belongs to type I at the macroscopic level. Besides, many minor cracks have been found on the cone top of seat bores at the discharge valve and the suction valve.

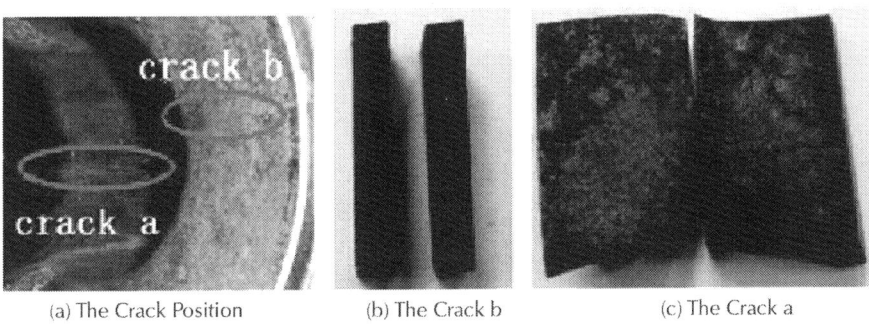

(a) The Crack Position   (b) The Crack b   (c) The Crack a

Figure 3: Macro-morphology for cracks.

The original fracture appearance of cracks are shown in Fig. 3(b and c). It can be observed that the fracture surfaces of the cracks were all covered by a thicker layer of brown or black deposit, which illustrated that the cross-section suffered serious oxidation after the cracks formed. What's more, the crack presented the brittle fracture mode for clear plastic deformation have not been found, and fracture relatively smooth as well as level off [14]. The another feature of the fluid cylinder's cracks is the thickly substrate material. It can easily be found by naked eyes, so penetrating break cases appear barely, which lead to the cross-section of fracture been made up of crack initiation and crack propagation, and with the characteristic of large fracture area because of length and depth.

The crack section has been treated with acetone and then cleaned and desiccated by ultrasonic dryer, shown in Fig. 4. There are a large number of radial lines on the section. The crack source area can be found if it goes along with the reverse radiate direction shown at the red[1] arrow place, which is very small. The most area occupied by crack propagation, about 95%. At the same time, it can clearly be found that the section is made up of two stages with crack initiation as well as crack propagation, namely, inexistence of transient breaking area. Macro-characteristics of shelly texture have been observed, which can prove the fatigue damage of fluid cylinders. The shelly texture demonstrated the intermittent growth of cracks which was caused by the discontinuous working state of frac-pump.

(a) Crack a   (b) Crack a

(c) Crack b

Figure 4: Macro-morphology for cracks section (67×).

The morphology of cleavage fracture can be clearly surveyed from Fig. 5, these splitting steps illustrated the fluid cylinder's fracture mechanism on the crack source areas, which caused by high circumference stress. In horizontal direction of crack spreading areas the typical fatigue striations have been observed (presented in Fig. 6). Fatigue steps and secondary cracks are observed on some of the fractures. Besides, the fatigue striation extended in brittle form.

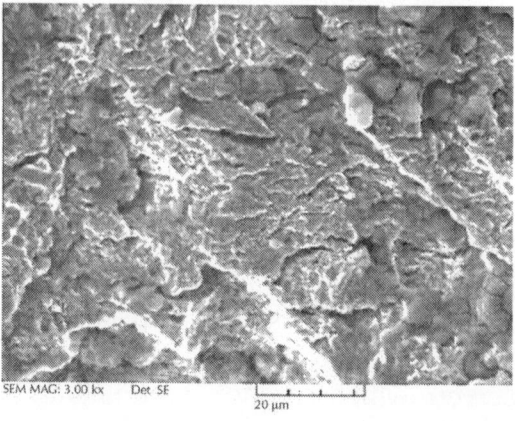

Figure 5: Crack 'a' initial area.

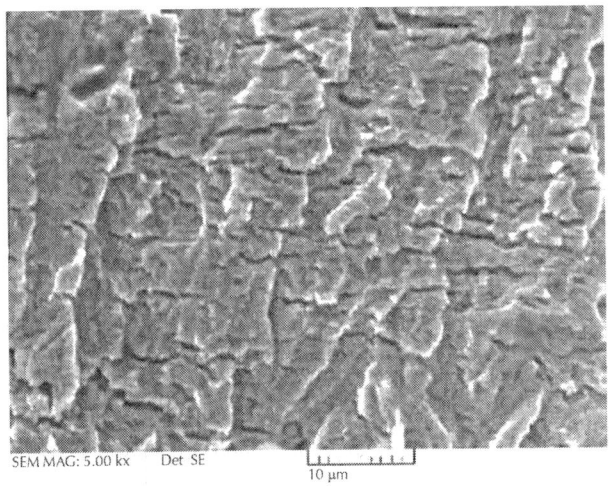

SEM MAG: 5.00 kx   Det SE
10 μm

Figure 6: Crack 'a' growth area.

Chemical Composition Analysis

The material of fluid cylinders is a high strength alloy steel, its material trademark is AISI-E4330, which has been quenching and tempering treatment. Chemical composition of the fluid cylinder was tested by using direct-reading spectrometer, listed in Table 2.

Table 2: Chemical compositions

Element mass per-cent	C	Si	Mn	P	S	Cr	Mo	Ni	V	Al	Cu
ZM-1#	0.30	0.25	0.79	0.008	0.001	1.57	0.58	2.44	0.198	0.008	0.126

The content of alloying element holds much, such as Cr, Mo, Ni as well as V, etc. It can relieve fluid cylinder's wear environment, corrosion and high working stress for longer service working life.

Metallographic Structure

Metallographic samples are cut from the longitudinal sections of the fluid cylinder near the fracture surface, which is observed by senior metallographic microscope (instrument model: OLYMPUS-PMG3) after the metallographic samples were prepared and etched by 4% nitric acid in alcohol [15]. The metallurgical microstructure is shown in Fig. 7 that the metallurgical microstructure is mainly composed of tempered sorbite and ferrolites contained in the fine needles.

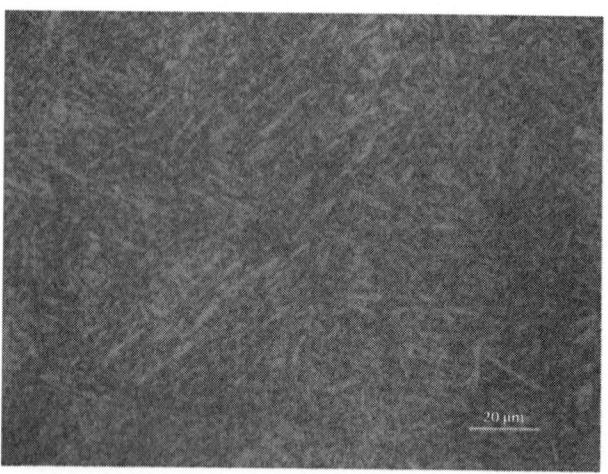

Figure 7: Metallographic structure (1000×).

Hardness Testing

Hardness testing is a simple and efficient approach to evaluate mechanical properties of material[16] and [17], In the box cavity near the surface along the radial direction, the hardness testing samples were cut out as shown in Fig. 8, which were observed by digital superficial Rockwell Hardness Tester (600MRD). Nine points had been tested at 5 mm interval along the radial direction. The test results are listed in Table 3. The data indicate that some points' value had not meet the technical requirements where closing to the surface for Valve box materials surface hardness requirement for 33–40 HRC after quenching and tempering treatment.

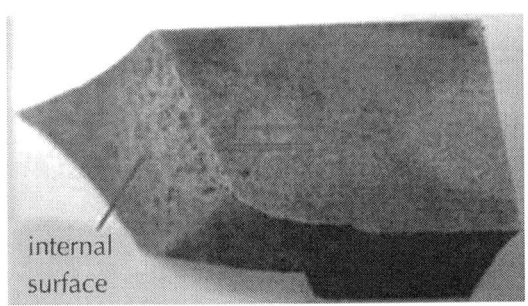

internal surface

Figure 8: The specimen for hardness test.

Table 3: The data of hardness test

Points value	1	2	3	4	5	6	7	8	9
ZM-1#	29.9	31.1	32.9	33.4	33.4	33.5	34.1	34.1	35.1

Mechanics Performance Test

Tensile test with five groups were performed by utilizing micro-control electronic universal material testing machine. Fatigue limit (σ^{-1} = 592 MPa) also obtained from the method of fatigue groups tested on rotating bending fatigue machine (PQ-6) in completely reversed bending. The testing data were shown in Table 4 and Table 5 and Fig. 9, which would be applied into fatigue prediction such as S–N or ε–N curve.

Table 4: Tensile test data

Material mark	Specimen numbers	Diameter d_0 (mm)		Cross sectional area A (mm²)		Gauge length L_0 (mm)		Elongation at break (%)	Reduction of area (%)	Yield strength (MPa)	Ultimate tension (kN)	Tensile strength (MPa)
		Before test	After test	Before test	After test	Before test	After test					
E4330	1#	12.71	8.40	123.64	55.39	50.00	59.11	18.22	55.20	982	131	1060
	2#	12.73	8.36	123.64	54.86	50.00	59.30	18.60	55.63	973	130	1053
	3#	12.72	8.39	123.25	55.13	50.00	59.22	18.44	55.36	970	129	1054
Average value		12.72	8.38	123.51	55.13	50.00	59.21	18.42	55.40	975	130	1056

Note: Modulus of elasticity: 2.06E5 MPa; Possion's ratio 0.28.

Table 5: Fatigue test data

Design test sequence	Actual test order	Design stress (MPa)	σ_a/σ_b	Stress ratio	Actual load (N)	Farmar weight (N)	Actual stress (Mpa)	Cycle-index	
1	2	624	0.5909	−1	222	155	628.1314395	2.919550E+05	Failure
				−1				3.88093E+05	Failure
2	6	612	0.5795	−1	217	150	613.984335	2.68024E+05	Failure
				−1				3.193060E+05	Failure
3	3	600	0.5682	−1	212	145	599.8372305	1.138070E+05	Failure
				−1				6.053590E+05	Failure
				−1				2.887790E+05	Failure
4	4	588	0.5568	−1	207	140	585.690126	1.085344E+07	Exceed
				−1				1.052623E+07	Exceed
				−1				1.179826E+07	Exceed

5		576	0.5455	−1	202	135	571.5430215	1.052665E +07	Exceed
				−1	202	135	571.5430215	1.024163E +07	Exceed
6	−	564	0.5341	−1	−	−	−	−	−
7	−	552	0.5227	−1	−	−	−	−	−
8	−	540	0.5114	−1	−	−	−	−	−
9	1	528	0.5	−1	187	120	529.101708	1.015492E +07	Exceed
				−1				1.019362E +07	Exceed

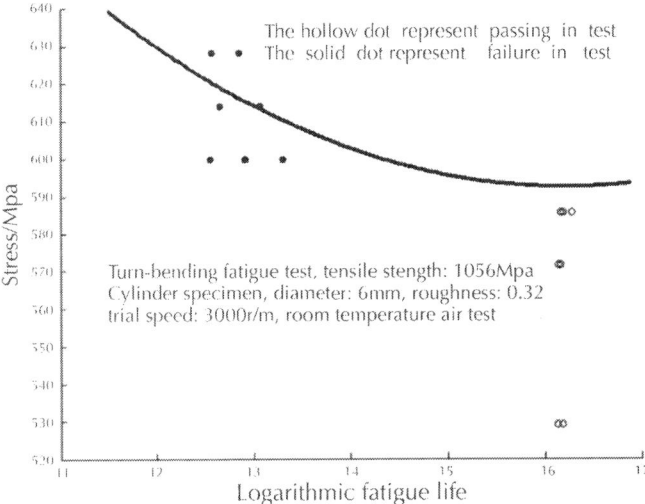

Figure 9: *S–N* curve.

FLUID CYLINDERS FATIGUE PREDICTIONS

Before the fatigue life prediction of fluid cylinders, the static strength analysis should be accomplished firstly. The boundary condition (Fig. 10) is fully-constrained the head face that closing to the power end case, and defined load cases for discharge thread bore, suction thread bore, the areas above discharge valve as well as the following areas of discharge valve respectively, the max inner working pressure is 137.9 MPa for fluid cylinders.

Figure 10: Loading and boundary condition.

The static strength analysis result indicates that the maximum Von Mises stress achieved 584 MPa. There are stress concentrations near the bore intersection, the cone top of seat bores of discharge valve and suction valve, the place between discharge bore and discharge 'valve lift' area, etc. Which were illustrated in Fig. 11. In the process of fatigue analysis, the frequency of fatigue load was defined as 0.000167 h according to pump's stroke of 100, and all cylinders is under working stress with phase difference 144°, in turn. The fatigue load is shown in Fig. 12. Making these areas discharge valve maintain normal pressure 137.9 MPa, while the following areas of discharge valve defined as single peak load with the maximum pressure of 137 MPa and minimum 0.3 MPa.

Figure 11: Von Mises stress.

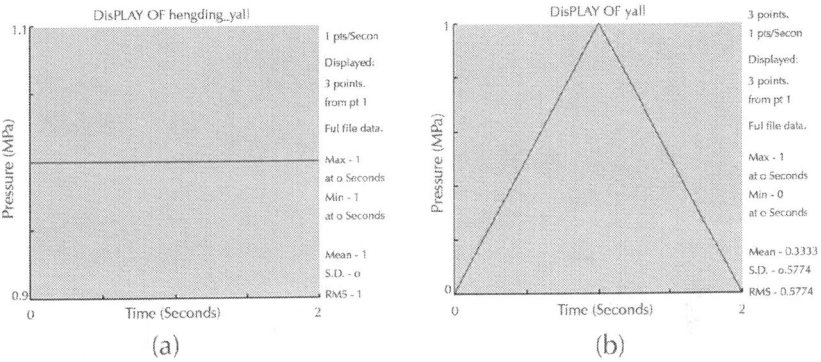

(a)   (b)

Figure 12: Fatigue load.

Total Life Analysis (S–*N*)

A metal subjected to a repeated load will breakdown at a stress level lower than that required to cause fracture on a single application of the load has been universally recognized since 1830. The nominal stress (*S–N*) method was the first approach which was developed to try to understand this failure process and was still widely used for simple fatigue prediction nowadays. Fatigue life cloud (Fig. 13) was calculated by Msc. fatigue software. From right to left, the cylinders are marked as 1#, 2#, 3# in this paper based on Fig. 13.

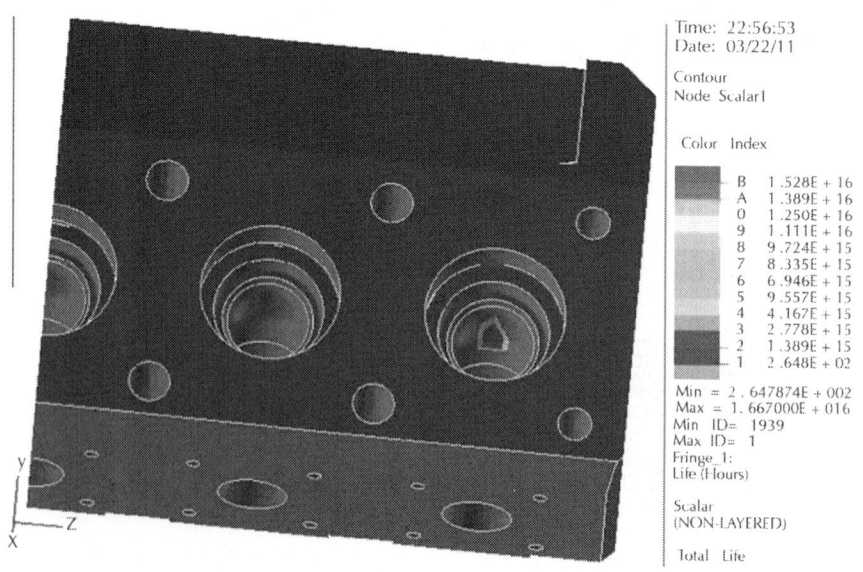

Figure 13: The result of S–N prediction.

From the prediction results, the fatigue life of fluid cylinders is generally sufficiency with the fatigue life 1.52E16 h and repeats 1E20, but there is hidden trouble of fatigue damage at specific bore intersection of 1# cylinder, whose fatigue life is 265 h and repeats 1.555E6.

The Strain–life Methodology

The S–N approach applied in where the stress is nominally within the elastic range of the material. From this point of view, the nominal stress approach is best suited to the area of the fatigue process which is known as high-cycle fatigue, so it does not work well in the low-cycle region where the applied strains have a significant plastic component. For fluid cylinders, a strain-based methodology should be processed. Fig. 14 is the simulation result of fatigue life for fluid cylinders on the basis of the strain–life methodology. Fatigue damage can be clearly observed at the region of bore intersection, cone top of seat bores for discharge valve and suction valve. Cutaway view of 1# cylinder reflects more specific image of fatigue weak areas (Fig. 14b), from which some information could be got that the cylinder began to produce cracks after serving rough 353 h and repeats 2.12E6. These were great accord with the result of failure analysis for fluid cylinders.

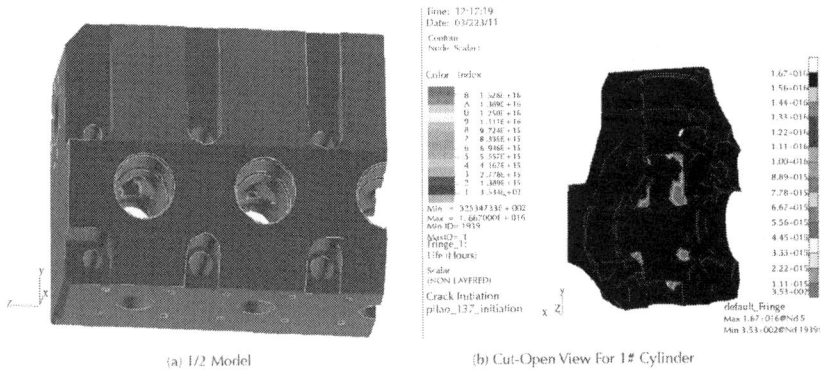

(a) 1/2 Model   (b) Cut-Open View For 1# Cylinder

Figure 14: The result of strain–life methodology (hours).

CONCLUSIONS

- The failure analysis of the scraped fluid cylinder illustrates that its fatigue cracking results from the major circumferential circularly stress. Owing to the corrosive of working medium, material performance will drop after a period of service, which will accelerate the crack forming and propagating.

- According to the results of the static strength analysis of fluid cylinders, the stress concentration of 1# cylinder is highest, 3# cylinder is the lowest. The stress distribution of the fluid cylinder gradually becomes fine from both sides to the middle; stress concentration appears at the intersection bore, discharge hole, the top of conical surface of discharge, suction valve seat bores and the fluting of inhaled bulkhead, but the maximum stress is lower than the yield stress. Generally, the structure of the fluid cylinder can meet the requirement of static strength.

- The results of fatigue prediction on the basis of S–N approach indicates that, the service life of fluid cylinders is about 265 h, fatigue weak area appears at the bore intersection of 1# chamber; the analysis result of crack initiation suggests that fatigue crack starts at the bore intersection as well as the top of conical surface of discharge and suction valve seat bores, where have large stress concentration.

- Comparison the results of failure analysis with the fatigue prediction, the crack initiation analytic approach conforms to the real damage, which explains that the response to cyclic loading is controlled by strain at the fatigue weak area for fluid cylinders' materials.

ACKNOWLEDGMENTS

The study was funded by national oil and gas major projects under Grant No. 2011ZX 05048-04HZ as well as natural science foundation of china under Grant No. 50905149.

REFERENCES

1.   Li Mo, Qing You Liu, Yun Huang, Jian Zhong Yang. The analysis of strength and life for fracturing pump fluid end. China Petrol Machinery 2009; 02:29–31.

2.   Xiang Yibing, Liu Yongming. Application of inverse first-order reliability method for probabilistic fatigue life prediction. Probab Eng Mech 2011; 26:148–56.

3.   Niesłony Adam, Kurek Andrzej, el Dsoki Chalid, Kaufmann Heinz. A study of compatibility between two classical fatigue

curve models based on some selected structural materials. Int J Fatigue 2012; 39:88–94.

4. Liu Y, Mahadevan S. Probabilistic fatigue life prediction using an equivalent initial flaw size distribution. Int J Fatigue 2009; 31(3):476–87.

5. Grooteman F. Adaptive radial-based importance sampling method for structural reliability. Struct Saf 2008; 30(6):533–42.

6. Minasny B, McBratney AB. A conditioned Latin hypercube method for sampling in the presence of ancillary information. Comput Geosci 2006; 32(9):1378–88.

7. Liu Y, Mahadevan S. Efficient methods for time-dependent fatigue reliability analysis. AIAA J 2009; 47(3):494–504.

8. Zhao Guozhen, Zhan Renrui, Tao Chunda. Explosive autofrettage for extra high tension units. China Petrol Machinery 1991; 19(9):1–9. 47.

9. Tao Chunda, Zhan Renrui, Han Lin. Research on fatigue strength of 100 MPa autofrettaged fluid pump end. J Mech Strength 2005; 27(1):104–7.

10. Shi Min, Zhou HouJun, Sun YuLong. The FEM analysis and fatigue prediction for autofrettage fluid cylinder. Petro Chem Equip 2009; 12:21–3.

11. Lei Zhang, Zheng Liang, Hua Jiang. Strength and life analysis of 140 MPa fluids pump end. Oil Field Equip 2011; 40(4):26–8.

12. He Xia, Liu QingYou, Zhao Min, Zhao Yong, Tian JiaYu. Fatigue prediction for pump end of high pressure fracturing pump[c]. Adv Mater Res 2011; 337:81–6.

13. Xiang Y, Lu Z, Liu Y. Crack growth-based fatigue life prediction using an equivalent initial flaw model. Part I: uniaxial loading. Int J Fatigue 2010; 32(2):341–9.

14. Rong Guo Zhao, Xi Yan Luo, et al. Analysis on fatigue damage and fractography of GH4133B superalloy used in turbine disk of aero-engine. J Mech Eng 2011; 47(6):92–100.

15. Zhang YiLiang, Jiang GongFeng, Xu XueDong, Ding DaWei. Stability and failure analysis of steering tie rod. J Beijing Univ Technol 2010; 36 (10):1317–23.

16. Yun Sun, Sheng Lai Wang, Qing Tian Gu, Jian Xu Ding, et al. Microhardness measurement of KDP crystal grown by point-seed technique. Funct Mater 2011; 10(42):1829–32.

17. Wei Guo Mao, Qiang Chen, Bin Zhang, Jie Wang. Investigation of elastic modulus and hardness of air plasma sprayed thermal barrier coatings. Mater Eng 2011; 10:66–77.

Chapter 5

Failure Analysis of Counter Shafts of a Centrifugal Pump

G Das[1], A.n Sinha[1], S.K.Mishra (Pathak)[1], and
D.K Bhattacharya[1]

[1]National Metallurgical Laboratory, Jamshedpur 831 007, India

ABSTRACT

An analysis of the premature failure of two counter shafts used in centrifugal pumps for lifting slurry has been carried out. Chemical analysis, microstructural characterisation, fractography, hardness measurement, tensile and Charpy impact tests were used for the analysis. The chemical compositions for the shafts were as per recommendation. The microstructure of one of the shafts was ferritic–pearlitic and its mechanical properties were inferior to the recommended values. For the other shaft the microstructure was tempered bainite; although the impact energy satisfied the specification, the other properties (hardness, UTS) were inferior. It was concluded that the improper heat treatment was the prime cause for the premature failure of the shafts.

INTRODUCTION

A shaft is a metal bar usually cylindrical in shape (solid or hollow), used to support rotating components or to transmit power or motion by rotary or axial movement. Shafts operate under a broad range of service conditions including various corrosive environments and a wide temperature range. Shafts may be subjected to a variety of loads such as tension, torsion, compression, bending or a combination of these. Shafts are also sometimes subjected to vibratory stress 1 and 2.

Shafts are made of various materials according to their applications and requirements. EN24 (AISI/SAE4340) steel is one of the common shaft materials. This is a medium carbon, low alloy steel. It is used where high strength and toughness are required for thick sections. It combines deep hardenability with ductility, toughness and strength 3 and 4. It also has good fatigue resistance. This steel can be case hardened without difficulty and finds many applications [5]. Its properties can be tailored by varying heat treatment schedules to get a good combination of mechanical properties and microstructure 6 and 7. Hardening can be done by oil quenching (up to 75 mm diameter) or by water quenching (for larger sections)[8]. After hardening by either process, tempering is carried out to reduce internal stresses and to optimise mechanical properties. Impact energy is one of several mechanical properties which is governed by the tempering treatment. It has been found that impact toughness may differ for various microstructures though their hardness value may be kept at a specific level [9].

This paper presents the analysis of failure of two EN-24 shafts used in centrifugal pumps for lifting slurries in a power plant. The failed shafts were made of EN24 steel. The shafts were of 6 cm (shaft A) and 7 cm (shaft B) diameter and were required to work for at least three years of continuous operation. Shaft A failed only after 16 days and shaft B after 10 days of operation. A schematic diagram of the entire pump is shown in Fig. 1 and the location of fracture is indicated by arrow marks. For both the cases, the breakage of the shafts was found to be in the centre of the pulley. The tensions of the belts were given as per the recommendation of the belt manufacturer. Prior to failure, no abnormal vibration was observed in the bearings associated with the shafts. Also after the shaft failure, the bearings were found to be in good condition. In Fig. 1 the motor, hydro-coupling and bearings 5 and 6 are mounted on one base whereas bearings, 7 and 8 and the pump are mounted on a separate base.

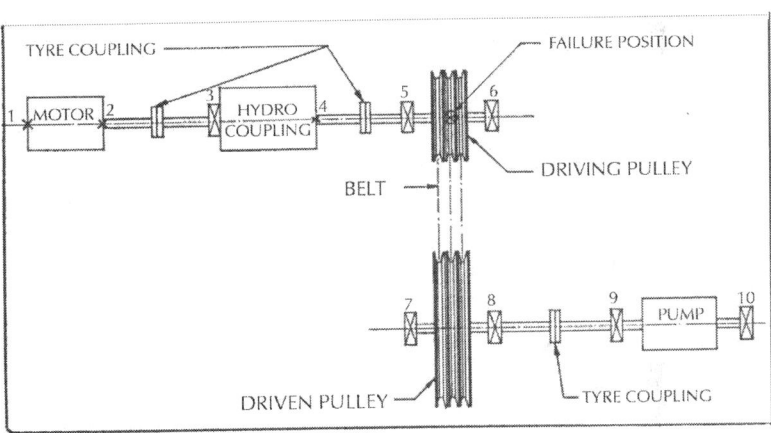

Figure 1: Schematic diagram of the failed centrifugal pump.

EXPERIMENTAL PROCEDURE

The microstructure of the shaft material was analysed by optical microscopes and a JEOL 840 scanning electron microscope (SEM) equipped with an energy dispersive X-ray (EDX) analysis facility. The composition of the shaft material was determined by using a standard spectrometer analyser as well as by using SEM–EDX. The samples for microstructural studies were prepared in the usual metallographic manner both in the transverse and longitudinal direction of the shaft axis. They were polished and etched with 5% nital (nitric acid in ethanol). Fractography of the broken shafts was carried out by SEM. Hardness testing was performed using a Vickers Hardness testing machine under 30 kg load. Tensile tests were carried out on cylindrical specimens as per ASTM standards by using a servo hydraulic INSTRON machine. Standard sized specimens for Charpy impact tests were made in longitudinal and transverse directions from both the shafts. The specimens were then tested in a Wolpert instrumented impact testing machine using 100 and 150 J hammers to get impact toughness values.

RESULTS

Visual Examination

Visual examination of the failed end gave the appearance that both the shafts failed by fatigue. A macroscopic view of the failed region of shaft B is shown in Fig. 2. Signs of smearing and distortion at the key edge were observed. Both the shafts failed at the end where the pulley is fitted as shown by the arrow mark in Fig. 1.

Figure 2: Macroscopic view of failed end of shaft B.

Chemical Analysis

The chemical analyses for both shafts A and B are listed in Table 1. SEM–EDX analysis also showed a similar composition for both shafts. It confirmed that they were made of EN24 steel.

Table 1: Chemical analysis for shaft material

Shaft	C	Si	Mn	S	P	Ni	Cr	Mo
A	0.41	0.25	0.47	—	—	0.97	0.99	0.12
B	0.44	0.20	0.43	—	—	0.96	0.85	0.22

Microstructural Analysis

Fig. 3 shows the representative microstructures in the longitudinal direction for shafts A and B. The microstructure in the transverse direction was essentially the same. The microstructure is observed as ferrite–pearlite for shaft A (Fig. 3(a)). Black patches indicate ferrite in the microstructure. A tempered bainitic microstructure was observed for shaft B, shown in Fig. 3(b). No significant inclusions or segregations were found; only a few pores were observed on the polished surface. The presence of pores are below the level of concern. The material used for the machining of the shafts was almost clean, only a few oxide inclusions of D4 fine ratings were observed.

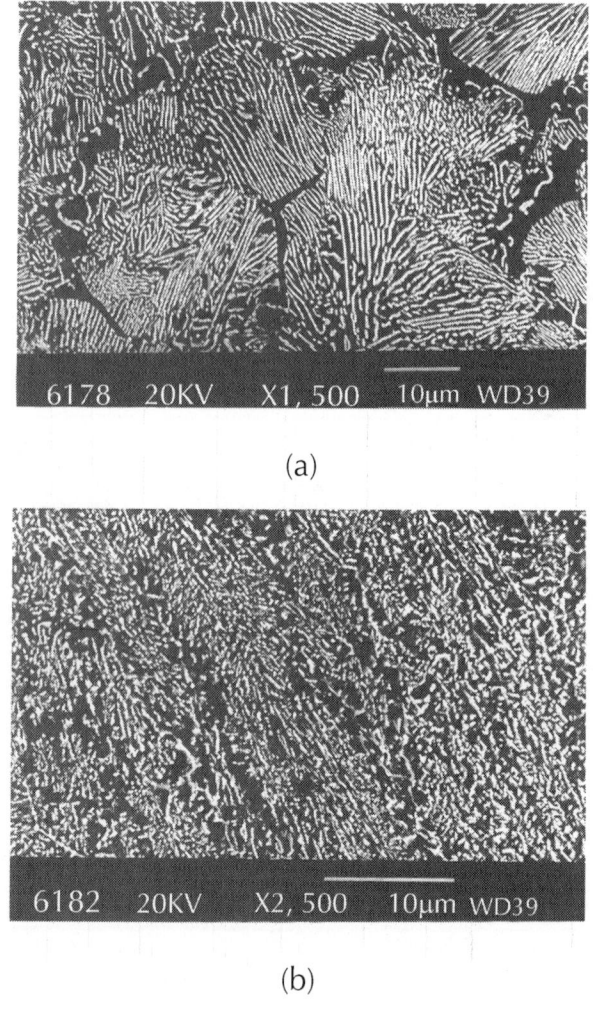

(a)

(b)

Figure 3: SEM micrograph of (a) shaft A, showing ferritic–pearlitic microstructure and (b) shaft B, showing tempered bainitic microstructure.

Fig. 4 shows the microstructure of the distorted region along the axis and along the transverse direction at the key region (indicated by * in Fig. 2) of shaft B. Lapping of thin layers and some smearing of metal was found near the key region (Fig. 4(a)). Microstructural analysis of Fig. 4(b) indicated the presence of plastic flow lines. It is also observed that the cracks had initiated from the smeared region (Fig. 4(b)).

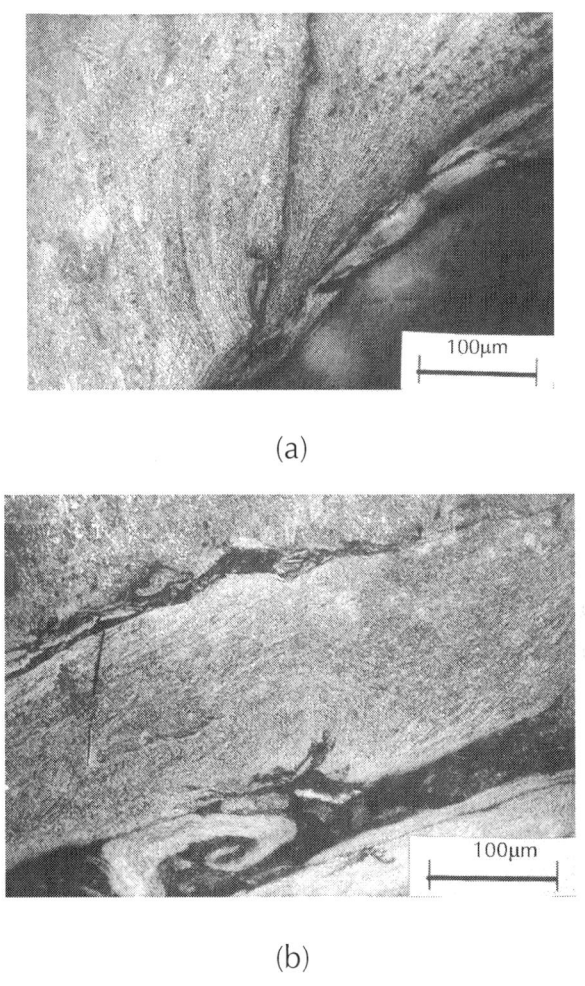

(a)

(b)

Figure 4: Optical micrographs for shaft B at key region (marked by * in Fig. 2) (a) along the axis and (b) along the transverse direction.

Fractography

An SEM fractograph for shaft A is shown in Fig. 5. The fracture surface of the broken shaft showed that the fracture was by fatigue (Fig. 5(a)) and the cracks initiated from the locking key sites. The apparent striations observed in Fig. 5(b) are actually likely to be fractured pearlite. Fig. 5(c) shows that the cracks had initiated from the keyway region. Propagation of secondary cracks was also observed, as shown in Fig. 5(d). A cavity of size approximately 300–400 μm was observed in the fracture surface.

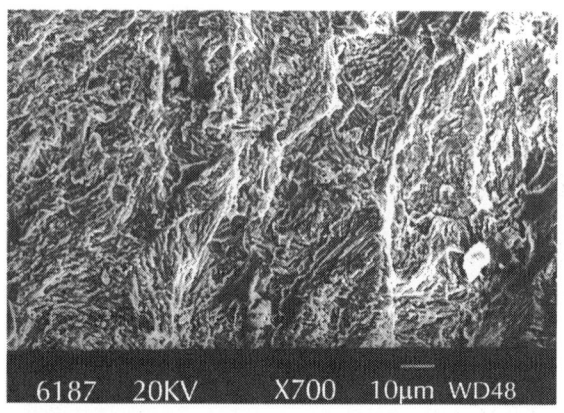

(a)

(b)

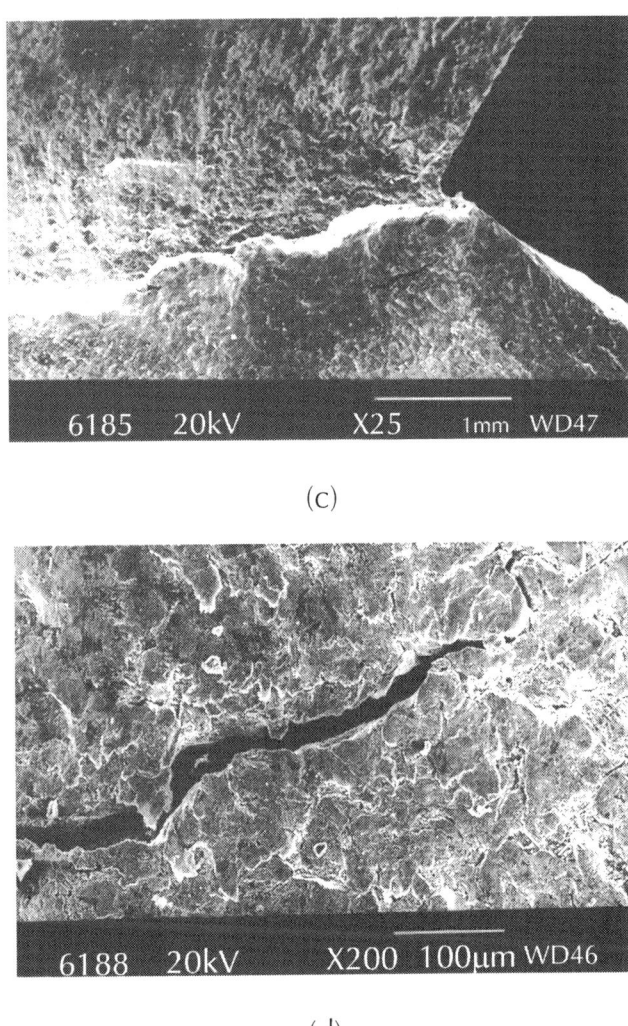

(c)

(d)

Figure 5: SEM fractograph for shaft A showing (a) fatigue failure, (b) apparent striations, (c) crack initiation region and (d) secondary cracks.

Similarly shaft B also failed by fatigue. The presence of striations in Fig. 6 supports the view of fatigue failure. Here also, the crack had initiated from the distorted key region where some plastic flow of metal was observed.

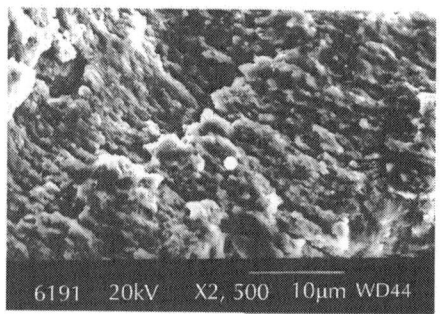

Figure 6: SEM fractograph for shaft B showing fatigue striations.

Mechanical Testing

Tensile

Cylindrical specimens for tensile tests were prepared from shaft B along the longitudinal direction. The test was performed with a strain rate of 10^{-3}/s. The stress–strain diagram is shown in Fig. 7. The yield strength, UTS and % elongation were found to be 530, 750 MPa and 22.55 respectively. The YS and UTS values were below the recommended level whereas % elongation satisfied the requirement.

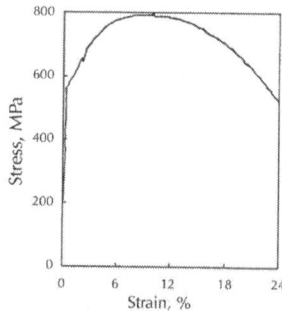

Figure 7: Stress–strain diagram for shaft B.

Impact

The results of the impact tests along the longitudinal and transverse directions are given in Table 2.

Table 2: Impact energy for both shafts

		Impact energy	
		J	ft lbf
Shaft	Specimen direction		
A	Longitudinal	17.5	12.96
	Transverse	15.5	11.1
B	Longitudinal	51	37.78
	Transverse	23	17

Hardness

Hardness values were obtained from the curved surface, as well as from the transverse cut surface. A number of readings at various locations across the cross section of the shafts were taken in order to determine the variation of hardness, if any. No appreciable variation in the hardness values was detected across the cross section. The average hardness values for shafts A and B were 170 and 240 HV respectively.

DISCUSSION

Shaft A

Chemical analysis showed that the shaft was made of EN24 steel which is a recommended material for such applications. The shaft was of a ferritic–pearlite nature, whereas for this type of shaft the final microstructure should be tempered martensite [5]. Proper tempering should be done after solution followed by oil quenching and rough machining. The impact energy obtained was far lower than the specified level. The hardness value was found to be 170 HV, which is also lower than the recommended value for such applications. According to specification, the hardness value should be 340–400 HV 4 and 5. Visual examination and fractographic study of the failed shaft disclosed that cracks had propagated from one of the keyways and failure was by fatigue. From the above discussions, it is clear that the heat treatment of the shaft was not properly done.

Shaft B

Shaft B was also made of EN24 steel, confirmed by chemical analysis. The microstructure was predominantly tempered bainite. The hardness value was found to be 240 HV, still below the recommended level, while the impact energy was as per recommendation (51 J) 4 and 5. Though the impact value satisfied the requirement for the present application, there was a lack of hardness and strength which might be a prime cause for the ultimate failure of shaft B.

Again, one of the keyways near the fracture surface was heavily deformed. Visual as well as microscopic analysis of the fracture shaft disclosed that the crack had initiated from the distorted region, shown in Fig. 1. Analysis of SEM fractographs confirmed the above. The microstructure of the plastically deformed region shows (Fig. 6) that the crack originated from the deformed region and propagated radially towards the centre. Here pulleys were assembled on the shaft by means of shrink fitting, which resulted in a stress raiser under bending stress. Improper fitting may also result in friction between the shaft and the pulley. Friction produces wear of the shaft, resulting in wear induced surface roughness and fretting, all of which might promote nucleation and growth of cracks. Friction can also activate metal flow as a result of plastic deformation.

CONCLUSIONS

The shafts (consisting of ferrite–pearlite for A and tempered bainite for B) were made of EN24 steel. The materials did not show significant inclusions or segregation. Only a few pores were noticed on the polished surfaces. Fractography observations revealed the signature of fatigue failure for both cases. The mechanical properties of the shaft materials were found to be inferior, though the impact energy for shaft B satisfied the requirements. Cracks were found to have originated from the key area of the shaft. For shaft A, improper heat treatment produced an undesirable microstructure and thus resulted in a low CVN toughness and low hardness of the shaft material. This was the primary cause of failure. For shaft B improper heat treatment resulted in low values of strength and hardness which made the material more prone to failure. Again, the shaft and pulley were not properly fitted, which led to fretting between the two components and aggravated the failure mechanism.

ACKNOWLEDGMENTS

The authors would like to thank Mr S. Das and Dr S. Ghosh Choudhary, for their help and many stimulating discussions. They are also grateful to Prof. P. Ramachandra Rao for encouragement and permission to publish this work.

REFERENCES

1.   Shaikh H, Kathak HS, Gnanamoorthy JB. Analysis of service water pump shaft failure. Prakt. Meta., 1990:27(7):362.

2.   Fraccis P. Shaft failure. Some common causes. Mach. Prod. Eng., 1974:124:195.

3.   Woolman J, Mottrum RA. The mechanical and physical properties of the British Standard EN Steel, vol. 2. Pergamon Press, Oxford, 1966.

4.   Agarwal V. Steel handbook. Gandhinagar: Vishwas Techno-Publishers, 1990.

5.   Metal handbook, failure analysis and prevention, vol. 11, 9th ed. ASM, 1986.

6.   Tomita Y. J. Mat. Sci., 1992;27(7):1705.

7.   Tomita Y. J. Mat. Sci., 1989;24(4):1357.

8.   Heat treaters guide, practice and procedure for iron and steels, 2nd ed. ASM International, 1995.

9.   Singh SR et al., editors. Proceedings of the clinic on failure analysis. India: NML, 1997.

Chapter 6

Failure Analysis of MVR (Machinery Vapor Recompressor) Impeller Blade

Tae-Gu Kim[a] and Hong-Chul Lee[b]

[a]Department of Safety Engineering, INJE University, Gimhae, Gyeongnam, 621-749, Republic of Korea
[b]Engine Division, ATRI(Aero-Tech Research Institute), ROKAF, Kumsa dong, Dong gu, Deagu, 701-799, Republic of Korea

ABSTRACT

This dissertation studies an accident resulting from the breakdown of an MVR impeller blade. Visual, stereoscopic and, SEM examinations have been carried out to find out the causes of the MVR blade defects. The results show that failure of the MVR blade was due to material casting defects. An initial crack started from casting defects where stress intensified and then, a fatigue crack progressed to the critical length of this crack. In addition to these defects, overstressing caused by overspeed accelerated the catastrophic failure of the blade.

INTRODUCTION

The compressor is one of the important pieces of equipment at a sugar factory where reduced productivity was experienced as a result of damage to the impeller blade. This paper emphasizes the importance of taking preventative measures through accident investigation.

The damaged cast impeller blade was used for 13 years and showed signs of a crack due to fatigue from exterior examination. It is thus assumed that there is a possibility of casting defects during manufacture and also, the change of operating conditions caused by external factors that effect rotating speed. Therefore, this paper will focus on the analysis of visual, stereoscopic, and SEM examinations to find out the exact causes of the impeller breakdown.

There are three identified causes from analysis of impellor accidents:

- The first cause is defects, such as avoids or cavities during the casting process [1] and [2].
- The second is increasing speed stimulated by the formation of cavity and rapid spread of rust [3].
- The third is rust created by the intensity of friction and the velocity of the moving fluid across each part [4].

NATURE OF ACCIDENT

The MVR, as a mechanical compressor, is used to increase the concentration of the refined liquid sugar for crystallization. The factories were supplied with power from two private electric generators (5 MW, 3 MW). The MVR was powered by the generators, and turned at 11,750 RPM. When the accident happened, there was a problem with the speed control system. The frequency of power supply increased from 60 Hz to 64 Hz, causing the speed of theimpeller to increase from 11,750 RPM to 12,530 RPM. As the result of that increase, the electric current and the vibration went up instantaneously, and the impeller blade broke. A picture of the broken impeller is shown in Fig. 1. The process at the factory is shown in Fig. 2 and the accident occurred during the concentration stage.

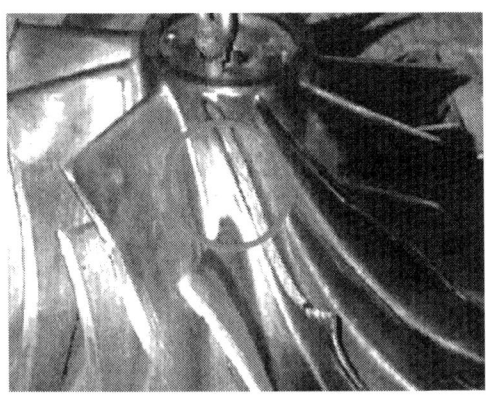

Figure 1: Photograph showing the failed MVR impeller blade.

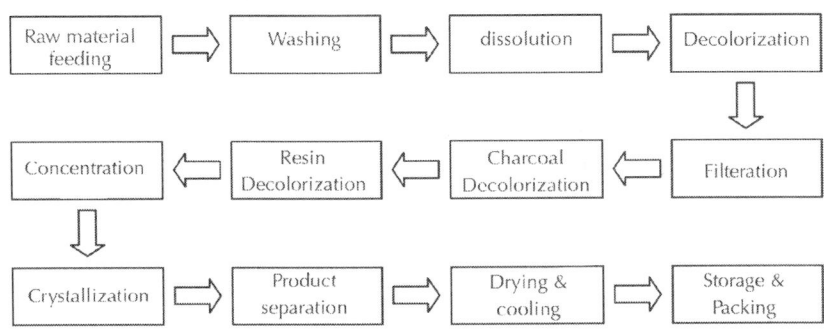

Figure 2: The workflow of the factory.

RESULTS OF ANALYSIS

Visual Examination

Beach marks as a broad distinctive feature are identified clearly just by visual observation. The fatigue crack progressed to 10.5 cm and then failed completely. As shown in Fig. 3, cracks existed on the fracture surface and several abrasive wear areas were considered as artificial damage after final separation or relative motion between the fracture surfaces. The chemical composition analysis of the failed impeller blade using the ICP (Inductively Coupled Plasma Spectroscopy) shows that it is similar to CB-7Cu-2 of cast stainless as shown in Table 1. There was not sufficient information about the material because the broken impeller was old and the model is no longer in production.

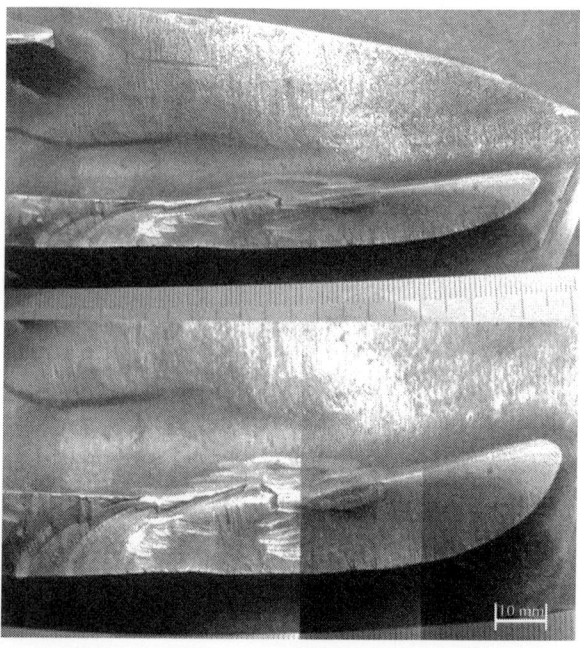

Figure 3: Visual examination of fracture surface.

Table 1: Chemical analysis of failed blade

Item	Composi-tion							
	C	Si	Mn	Ni	Cr	Cu	Nb	Fe
CB-7Cu-2	0.07	1.00	0.70	5.00~7.00	14.0~15.5	2.5~3.2	0.2~0.35	Other
Blade	0.06	0.34	0.44	5.21	12.67	1.33	0.21	Other

Stereoscopic Examination

From a wide-angle view, observation of the fracture surface permits it to be classified very clearly into two parts:

- The first is a rough and brittle fracture surface
- The second is a semicircle of the beach marks.

When the semicircle of the beach marks enlarged, it is easy to identify a 1.5 mm initial crack from which the crack propagated, Fig. 4.

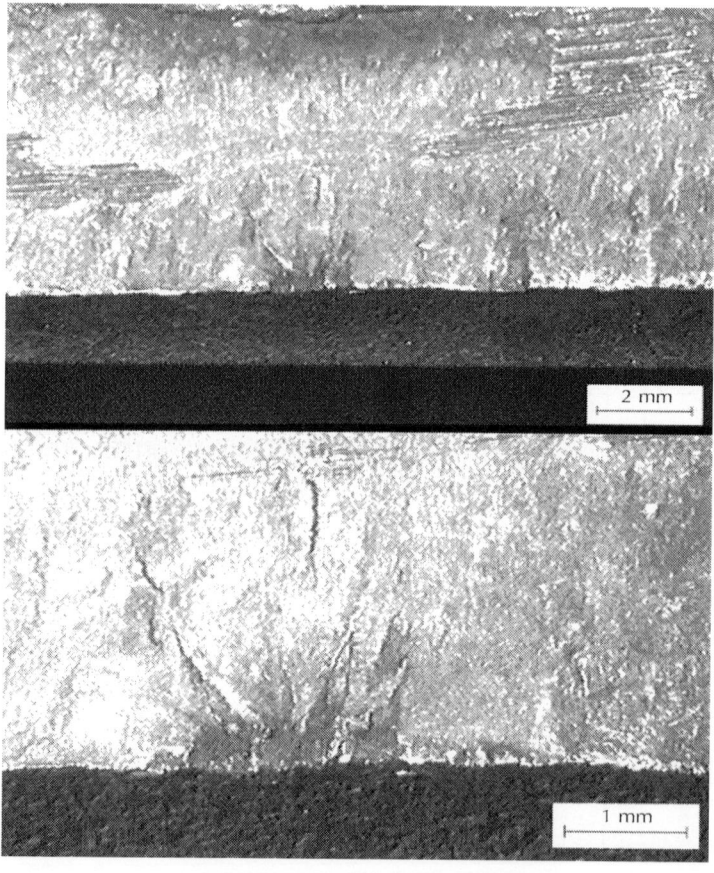

Figure 4: Stereoscopic examination of initial flaw and fatigue fracture.

SEM Examination

Fig. 5 shows the morphology of the fracture surface through the use of SEM to find the cause of the initial crack. As seen in Fig. 6, the fracture surface created by the fatigue crack is smooth, but the initial crack looks coarse (Fig. 7).

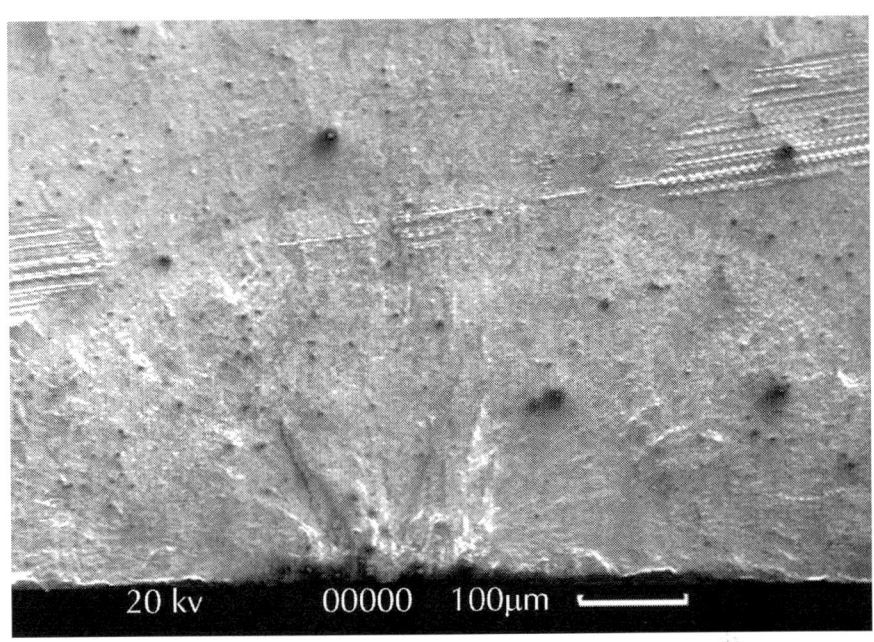

Figure 5: SEM micrograph showing appearance of fracture origin.

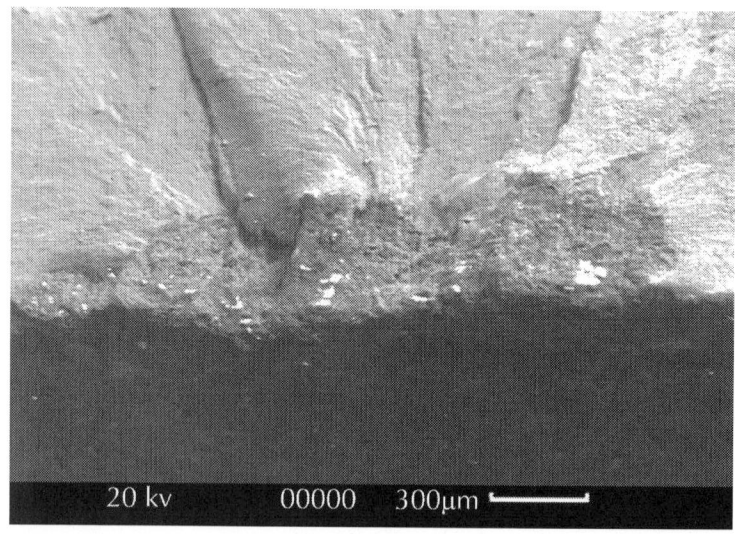

Figure 6: SEM micrograph showing difference between origin and fatigue fracture.

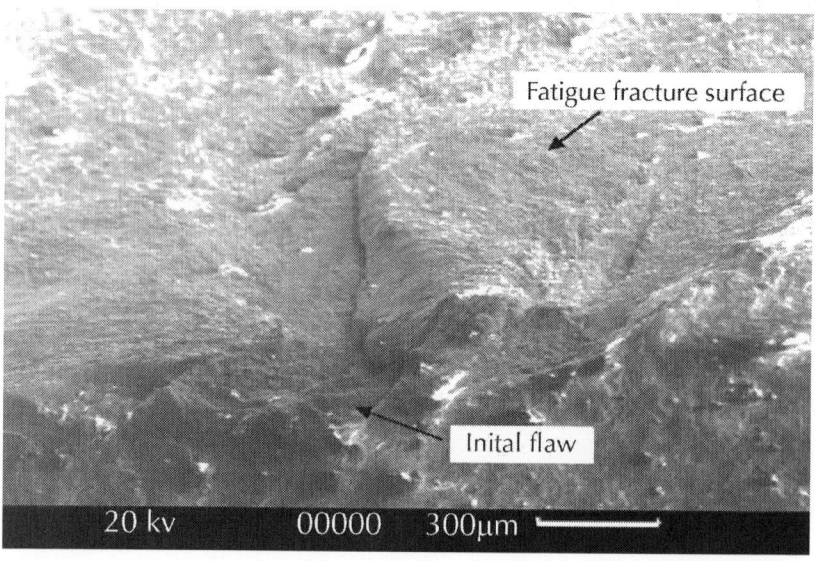

Figure 7: SEM micrograph showing initial flaw nucleated by surface defects.

The side of the initial crack has a surface crack that is considered to exist before the fatigue crack. EDS analysis of part of the initial flaw and fatigue fracture surface (Fig. 8) shows that there is a big difference in the quantity of the elements C, O and F, which are compared to the corrosive oxide Na and Ca which do not exist on the fatigue fracture surface, but are detected in the initial flaw.

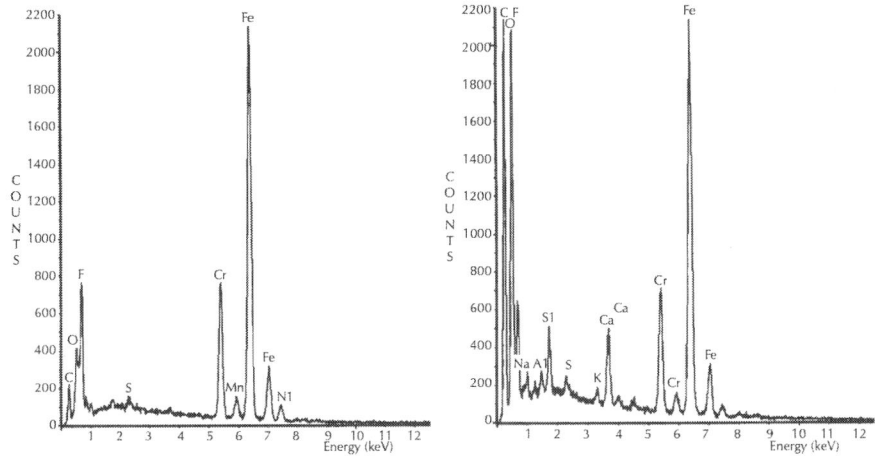

Figure 8: EDS spectrum analysis of initial flaw (left) and fatigue fracture surface (right).

There are lots of surface cracks, as seen in Fig. 9 and Fig. 10 that were discovered on the fracture surface of the impeller blade, which occurred during the casting process. The nucleation of the fatigue crack in the impeller blade is considered the result of casting defects. The surface crack became the initial crack and then, as stress intensified on this area, a fatigue crack progressed into a critical crack. In addition to these defects, overstressing from overspeed of the impeller accelerated the catastrophic failure of the MVR impeller blade.

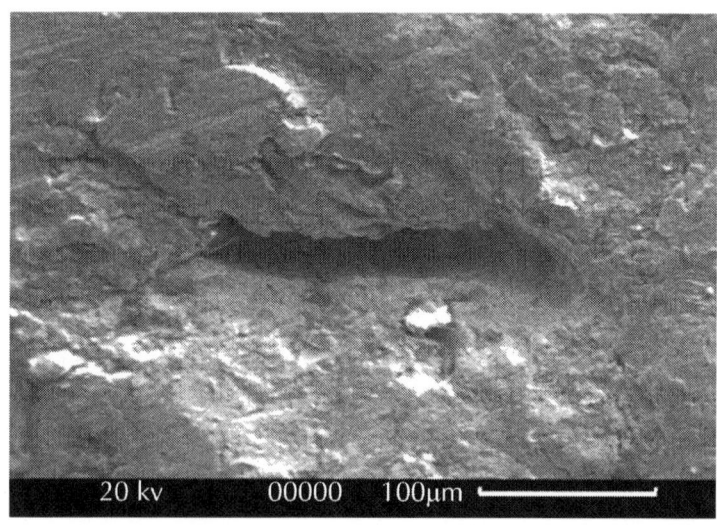

Figure 9: SEM micrograph showing material defects (cavity) in the surface of the failed component.

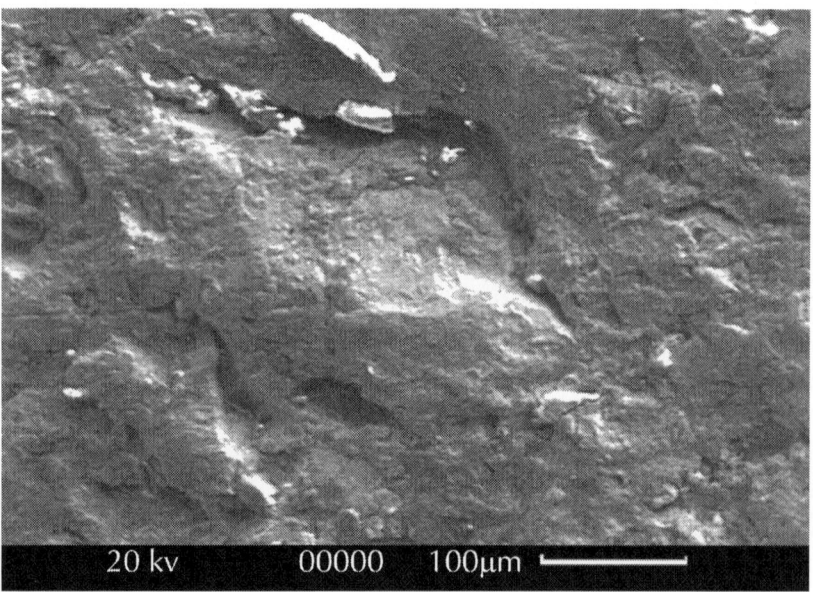

Figure 10: SEM micrograph showing material defects (pores, cavity) in the surface of the failed component.

DISCUSSION

No fault was found in the MVR involved in the accident through Non-Destructive Testing (NDT) of the blade during factory overall. NDT on the blade was carried out 2 weeks before the accident. The defect could not be detected through NDT. It is assumed that the accident happened from a fatigue crack through continued operation and changes in operational conditions.

Observation of the polished surface to check the material for porosity, which is easy to find in castings, is shown in Fig. 11. Fig. 12 shows the results of observation of the columnar structure. It confirms that the defect occurred when the impeller blade was produced.

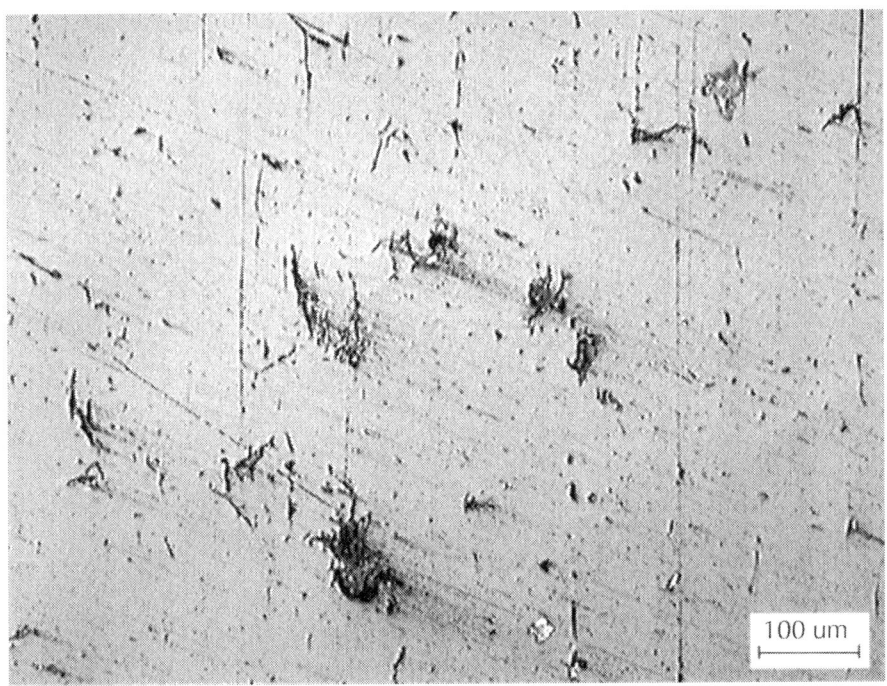

Figure 11: Metallography showing porosity found in the material.

500 um

Figure 12: Metallography showing the columnar structure.

There is a need for improvements in the production technique of the impeller blade in order that cracks are avoided during the casting process, and a close examination is necessary to find fatigue cracks. When regular inspections are conducted it is necessary to evaluate the accuracy of the examination carefully. It is also necessary to make a close examination of the crack on the fragile part to prevent the recurrence of similar accidents. It is suggested that providing vibration recording monitoring and an auto trip system would be a sensible precaution

CONCLUSIONS

From the investigation and analysis on the failure of MVR impeller blade, the conclusion can be summarized as follows;

- First, the initial crack in the MVR impeller blade was nucleated by the material-casting defect after a prolonged period of surface crack prorogation.

- Second, it is assumed that the MVR fractured due to a fatigue crack after a long period of peak operation (about 13 years).
- Third, overheating caused by the over-speed of the impeller blade in the MVR due to an increase in the oscillation frequency, is considered to be the final cause of the MVR breakdown.

ACKNOWLEDGMENTS

This work was supported by the 2002 Inje University research grant.

REFERENCES

- van Bennekom A, Berndt F, Rassool MN. Pump impeller failures—a compendium of case studies. Engineering Failure Analysis 2001;8 (2):145–56.
- Colangelo VJ, Heiser FA. Analysis of metallurgical failures. New York: Wiley; 1974.
- AimingF, JinmingL, Ziyun T. Failure analysis of the impeller of a slurry pump subjected to corrosive wear. Wear 1995;181-183(2):876–82.
- Prakash O, Pandey RK. Failure analysis of the impellers of a feed pump. Engineering Failure Analysis 1996;3(1):45–52.

Chapter 7

Failure Analysis of Monel Packing in Atmospheric Distillation Tower under the Service in the Presence of Corrosive Gases

Amir Mostafaei, Seyed Majid Peighambari, and
Farzad Nasirpouri

Laboratory of Failure Analysis, Research Centre for Advanced Materials and Minerals, Department of Materials Engineering, Sahand University of Technology, Tabriz, Iran

ABSTRACT

We have performed a failure analysis on a monel packing material subjected to sulfidation and sulfide stress cracking after a short period of service in an atmospheric distillation tower containing corrosive gases. Optical macroscopic inspections show that the failed packing material was basically covered with dark scales and corrosion products. Energy dispersive X-ray spectroscopy and X-ray diffraction spectrum show that the corrosion products contain sulfur as the main element along with oxygen indicating that oxidation may have been taken place beside sulfidation. Metallography of the failed samples reveals that cracks exist and were initiated on the surface from defects most likely developed due to the formation of porous and brittle sulfide layer accomplished by hydrogen embrittlement. Although nickel base alloys are an appropriate candidate owing to their high corrosion resistant in corrosive atmospheres, however, presence of H_2S, HCl, O_2 and naphthenic acid in crude oil lead to reduction in corrosion resistant of the monel 400 alloy.

INTRODUCTION

Corrosion is a commonplace issue in metallic structural constructions. It was reported that the cost of corrosion for an industrialized nation is close to 4% of gross national product (GNP) and around 8% was attributed to the oil and chemical industry. In ideal conditions, corrosion is predictable but due to different manufacturing history of alloys and the nature of the design, they exhibit completely different performances under the service [1] and [2].

In general, the petroleum industry consists of different units and lots of processes are utilized there. Since there are wide ranges of corrosive chemical compounds used during production and refinement of oil and other petroleum products, structural materials in service suffer severe corrosion conditions. Atmospheric distillation tower is a vital part in oil and petroleum industry where light and heavy chemicals compounds are separated into different products. High temperature corrosion in distillation units is a main concern to the refining industry. The presence of different compounds such as naphthenic acid and sulfur compounds significantly increases corrosion in the high temperature parts of the distillation units and equipments failure have become an important issue. Since different processes occur in each part of oil refinery constructions, the problem of correlating corrosion of a unit to a certain type of crude oil exists.

Atmospheric distillation tower usually consists of different separators called tray or packing. If diameter of tower is small (less than 60 cm), packing is a prevalent equipment. In contrast, tray is used in towers with bigger diameters (more than 60 cm). They are usually made of steel or nickel base alloys such as AISI 316L and monel 400 due to their appropriate corrosion resistance in high temperatures. In addition to the presence of hydrogen and carbon, some impurities such as ammoniac, metallic salts and sulfurous compounds are present. Among them, the latter has a negative effect on the metallic structures and develops corrosion and destruction of parts [3].

As has been reported, sulfur is a common corrosive impurity at high temperatures. When combustion occurs with an excessive amount of air, sulfur reacts with oxygen to form SO_2 and SO_3 creating an oxidative atmosphere containing $SO_2 + O_2$. This atmosphere is more corrosive than O_2 atmosphere in nature. Additionally, if H_2S exists in the atmospheric tower, it will be more corrosive than the oxidized atmosphere (SO_2). Thus, it motivates metals to form oxide and sulfide layers on their surface, which is porous and not protective [4], [5], [6], [7], [8], [9] and [10]. Based on the temperature, various corrosion compounds form on the surface of a metal. Furthermore, at high temperature, the presence of naphthenic acids may increase the severity of sulfidation, because these organic acids damage the sulfide film and consequently encourages sulfidation on alloys that would normally be expected to resist against this form of attack [11].

Nickel based alloys have been widely used in different constructions in petroleum, sea water environment and aerospace industries [12], [13], [14], [15], [16] and [17], due to it high corrosion resistant in different conditions. A wide range of elements are added to nickel base alloy to achieve specific properties such as mechanical and physical properties and higher corrosion resistance. Monel 400 is a precipitation alloy of nickel containing considerable amount of copper (approximately 30–33 wt% copper) as a substantial element to provide solid solution strengthening. Also, small amount of iron is added to improve the resistance of the alloy to cavitation and erosion in condenser tube application [1]. It was reported that nickel base alloy such as monel 400, is susceptible to sulfide stress cracking (SSC) when the surrounding environment contains sulfurous compounds such as hydrogen sulfide (H_2S), mercury salts and hydrofluoric acid. It should be considered that alloys with a modified composition still can be susceptible to mixed oxidizing–sulfidizing attack [18], [19], [20] and [21]. Hydrogen induced cracking (HIC) has been observed in monel alloys and intergranular crack can be observed after short time. Due to high weight percent of nickel in monel alloy (around 65 wt%), they are immune to chloride stress corrosion cracking (CSCC). Different parameters affect oxidation and sulfidation of nickel alloys classified into two groups: (1) environmental effects that include concentration of hydrogen sulfide, temperature and applied potential and (2) variations in physical, mechanical and metallurgical properties of alloy during production procedures [1], [22], [23] and [24]. Thus, material selection is a critical stage in choosing a proper alloy for specific applications.

In this study, failure analysis of a packing material utilized in the atmospheric distillation tower was performed and the prominent mechanism of corrosion has been proposed. Finally, some tips are recommended to prevent or postpone this kind of corrosion.

MATERIAL AND METHODS

Failed and intact packings were achieved from a petroleum industry. They were used in atmospheric distillation tower at the temperature of 240–270 °C to separate liquid petroleum compound and gases from each other. The gases consist of a wide range of chemicals such as H_2S, HCl, ammoniac and C_6–C_7 which are extremely corrosive.

The chemical composition of the intact packing material was analyzed and is summarized in Table 1 which identifies it as Monel 400.

Table 1: Elemental chemical composition of the intact packing

Component	Ni	Cu	Fe	Mn
wt%	66.92	31.67	1.07	0.34

Metallographic examinations were performed on selected regions from cross sections of intact packing and failed one mounted and polished and chemically etched in nital 3% solution. The surface morphology and chemical composition of corrosion products were investigated using scanning electron microscopy (SEM- MV2300 Cam Scan) and energy dispersive X-ray spectroscopy (EDS-Oxford instruments). Additionally, X-ray diffraction (XRD) pattern of the intact and failed samples were recorded using D8 Advance-Bruckers AXS diffractometer with Cu Kα radiation source ($\lambda = 1.54$ Å) operated at 40 kV and 40 mA in the 2θ range 10–100° at the scan rate of 0.05° per second.

RESULTS AND DISCUSSIONS

Visual Inspection

Images of as-received specimens are shown in Fig. 1. As it can be seen, the intact packing is bright and free of any contamination. Fig 1b shows the failed packing sample which is brittle and contains corrosion products formed on the surface. Initial observation revealed that there were non-adhesive dark products on the both surfaces of the packing which were frail and easily broken to small segments. Visual inspection was performed and the results revealed that a thick black layer of corrosion products exists on the surface of packing after service. The failed sample was too fragile and it was easily broken to small pieces by loading a low force into it (Fig. 1c).

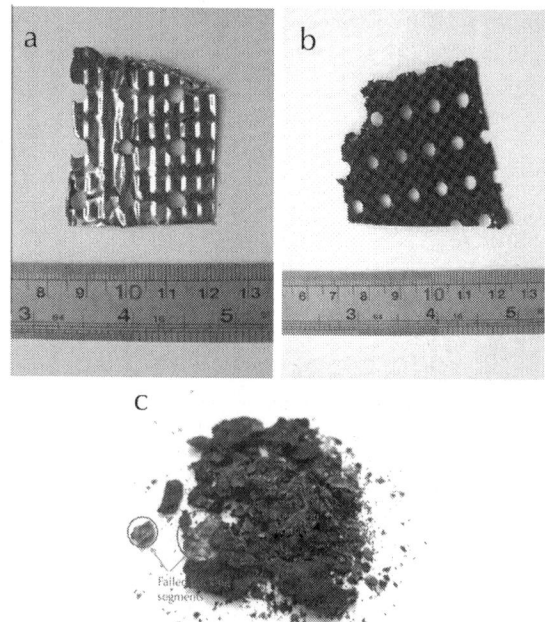

Figure 1: An image of the packing taken in the laboratory: (a) intact sample before use, (b) sample under the service and (c) failed packing containing packing segments and brittle corrosion products.

Metallographical Examinations

Morphology and microstructure of the intact and failed packing samples were investigated by optical microscopy. Fig. 2 shows optical images taken from the cross sectional part of the intact packing. Fig. 3illustrates optical micrographs taken from cross section of the failed sample. After polishing and etching procedures, metallographical examinations reveal a uniform morphology of two distinct phases in the intact sample (Fig. 2c and d). In addition, no crack is observed near both surfaces of the sample, while, intergranular micro-cracks are observed in the case of failed specimen (Fig. 3c and d); they are present on the both surfaces and grow along grain boundaries.

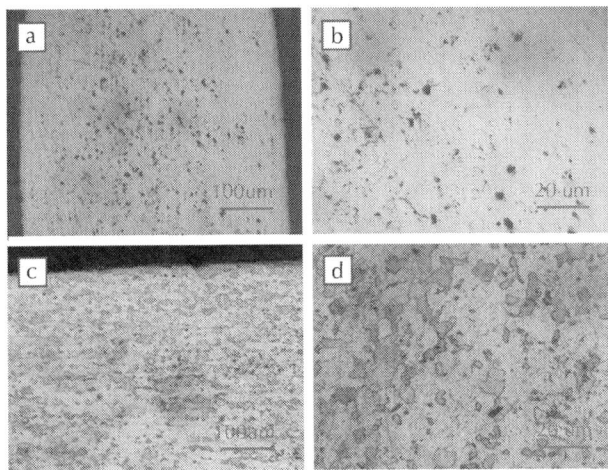

Figure 2: Optical micrographs of the intact monel 400 packing (a and b) before etch and (c and d) after etch in Nital 3%.

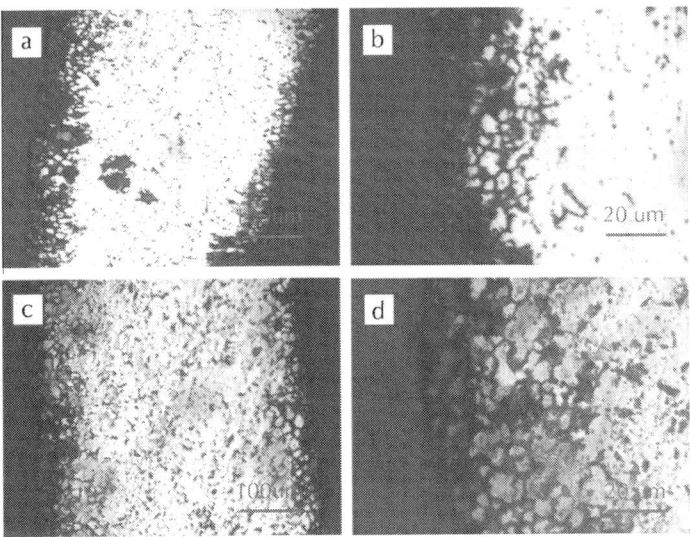

Figure 3: Optical micrographs of the failed monel 400 packing (a and b) before etch and (c and d) after etch in Nital 3%.

SEM Studies

Scanning electron microscopy was used to study the morphology of corrosion products. Fig. 4 shows SEM micrographs taken from failed packing specimen. Fig. 4a shows a typical SEM images taken from the cross section of the failed packing and Fig. 4b illustrates the morphology of the corrosion products formed on the surface of the failed sample. Based on Fig. 5a and b, it can be easily seen that corrosion products are not formed uniformly on the surface of the packing and a few deep and continuous cracks are present there.Fig. 5c and d shows a close look at the morphology of the corrosion products in two different magnifications. It is realized that the corrosion product has a porous structure on the surface.

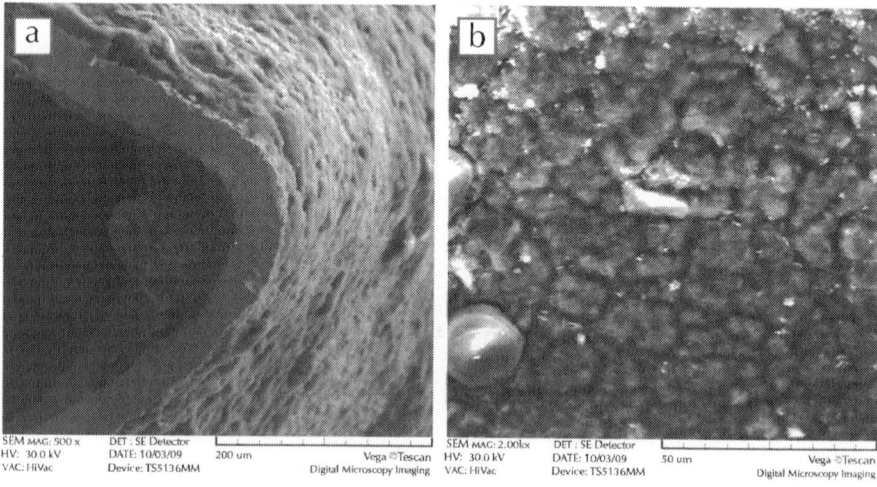

Figure 4: SEM images of the failed packing taken from (a) cross section and (b) surface of the failed sample covered with corrosion products.

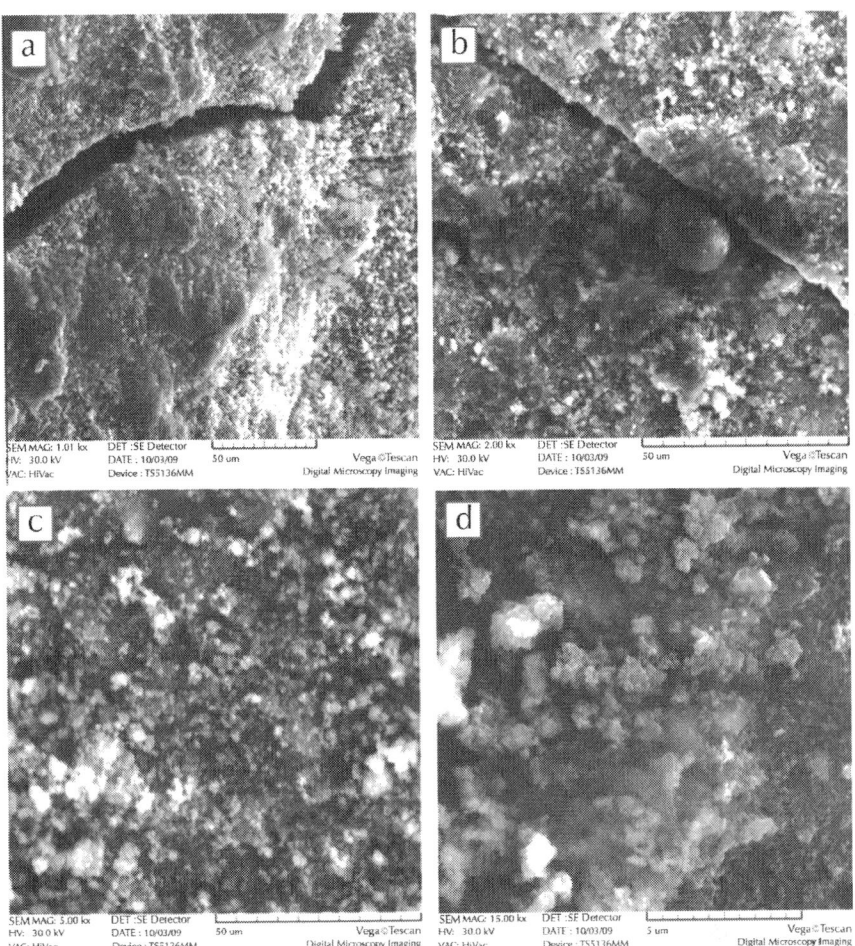

Figure 5: SEM images taken from surface of the failed packing sample in (a) 1000×, (b) 2000×, (c) 5000× and (d) 15,000× magnification.

XRD Examination

Qualitative analysis of the intact alloy and corrosion products was carried out by XRD. Fig. 6a shows the X-ray diffraction pattern of the monel alloy and Fig. 6b illustrates XRD pattern of the corrosion products. It is clear that the intact packing mostly consists of nickel and copper, while the failed sample consists of sulfides and oxides compounds such as NiS, Ni_2S_3, Ni_3S_4, Ni_7S_6, NiO, CuS, Cu_2S, CuS_2 and CuO. Additionally, due to presence of iron in the composition of the alloy, X-ray diffraction peaks of iron compounds are detected (such as $FeSO_3$).

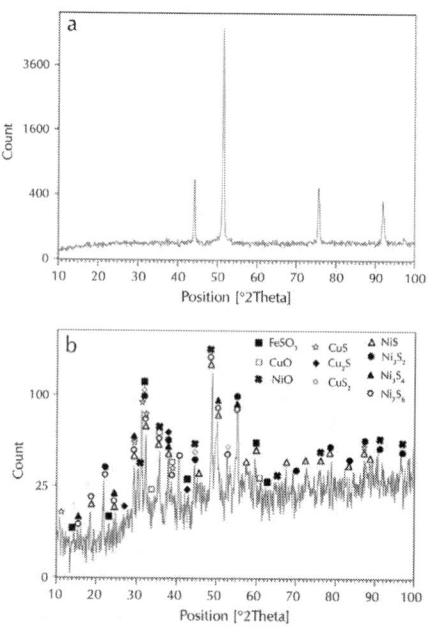

Figure 6: X-ray diffraction pattern of the (a) intact monel packing and (b) failed sample.

Chemical Analysis

Energy dispersive X-ray spectroscopy analysis (EDS) reveals that sulfur and oxygen exist in the corrosion products. Fig. 7A shows a SEM image of the failed sample and Fig. 7B illustrates different points on it with higher magnification. EDS elemental analysis results for these points are given in Table 2, Table 3, Table 4 and Table 5. As it can be seen, corrosion products consist of Ni, Cu, S, O and Fe. Additionally, Fig. 8 illustrates the EDS scan-line analysis of Ni, Cu,

S and O taken from cross section of the failed sample. As it can be seen from Fig. 8B, the existence of O and S peaks in EDS spectrum clearly demonstrates the oxidation and sulfidation of the monel under service. Most probably, a combination of nickel and copper in packing sample with sulfur and oxygen, present in hot corrosive gases in the atmospheric distillation tower, produces nickel and copper sulfides, oxides and other compounds under the condition of service. Since these corrosion products are not adhesive and compact enough on the surface of substrate, therefore, we cannot expect a complete protection of packing under severe corrosive condition.

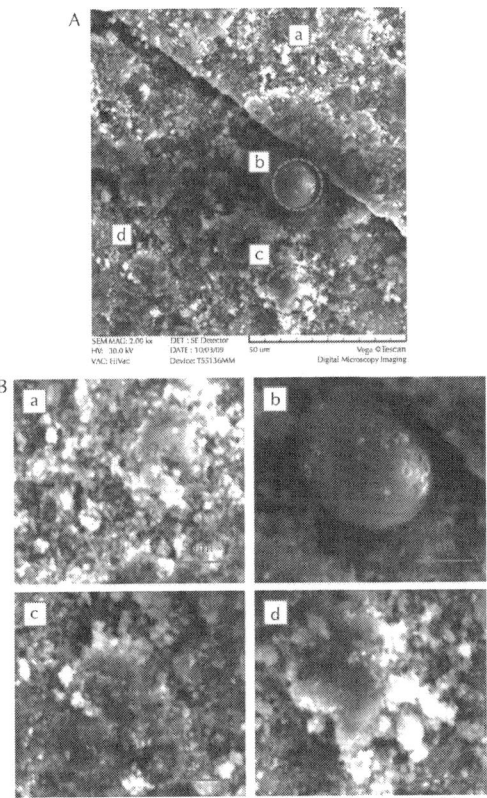

Figure 7: (A) SEM image of the failed packing taken from surface for EDS elemental analysis of the corrosion product in various points and (B) four selected points for EDS elemental analysis.

Table 2: EDS elemental analysis of the corrosion product taken from point 'a'

Element	App	Intensity	Weight (%)	Weight (%)	Atomic (%)
	Conc.	Corrn.		Sigma	
O K	6.96	0.7632	22.55	0.72	47.48
S K	4.82	0.6957	17.13	0.33	17.99
Mn K	0.76	1.0264	1.83	0.16	1.12
Fe K	0.39	1.1451	0.84	0.17	0.50
Ni K	20.32	0.9339	53.71	0.62	30.81
Cu K	1.43	0.8948	3.95	0.30	2.09
Total			100.00		

Table 3: EDS elemental analysis of the corrosion product taken from point 'b'

Element	App	Intensity	Weight (%)	Weight (%)	Atomic (%)
	Conc.	Corrn.		Sigma	
O K	3.05	0.5838	11.24	0.70	26.68
S K	10.55	0.7313	31.00	0.41	36.71
Fe K	5.31	1.0430	10.94	0.26	7.44
Ni K	10.32	0.9200	24.09	0.39	15.58
Cu K	9.39	0.8867	22.73	0.40	13.59
Total			100.00		

Table 4: EDS elemental analysis of the corrosion product taken from point 'c'

Element	App	Intensity	Weight (%)	Weight (%)	Atomic (%)
	Conc.	Corrn.		Sigma	
O K	0.40	0.5877	9.23	1.50	22.96
S K	1.52	0.7193	29.06	0.95	36.06
Fe K	0.97	1.0536	12.59	0.69	8.97

Ni K	1.65	0.9241	24.43	0.96	16.56
Cu K	1.61	0.8913	24.68	1.02	15.45
Total			100.00		

Table 5: EDS elemental analysis of the corrosion product taken from point 'd'

Element	App	Intensity	Weight (%)	Weight (%)	Atomic (%)
	Conc.	Corrn.		Sigma	
O K	1.88	0.5935	8.38	0.69	21.35
S K	7.08	0.7154	26.23	0.41	33.35
Cl K	0.48	0.5300	2.38	0.23	2.73
Fe K	5.54	1.0512	13.96	0.32	10.19
Ni K	5.91	0.9254	16.92	0.38	11.75
Cu K	10.83	0.8933	32.14	0.49	20.62
Total			100.00		

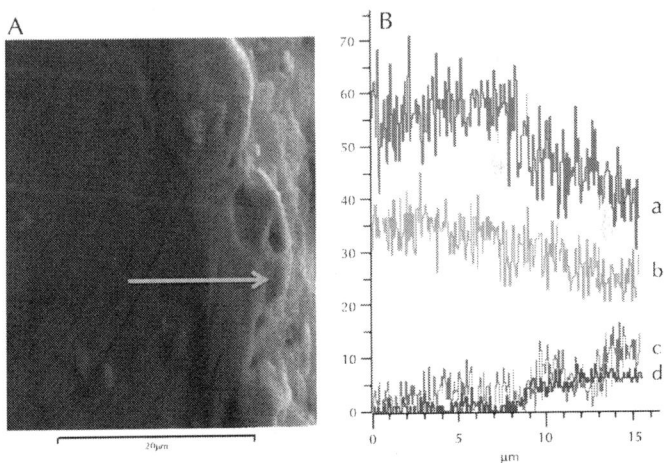

Figure 8: (A) SEM image of the failed packing taken from cross section and the direction of the EDS line analysis and (B) the results of the EDS line analysis of (a) Ni, (b) Cu, (c) S and (d) O.

Sulfidation and Oxidation Mechanisms in Nickel–Copper Alloys

In atmospheric distillation tower, high amount of sulfur compounds especially H_2S are present and because of its corrosive identity, sulfide scales are formed on the surface or beneath the bare metal. In general, nickel-base alloys are resistant under sulfidation or oxidation atmosphere, but performance of them in mixed sulfidation–oxidation atmospheres are completely different. In addition, the presence of other corrosive gases such as HCl and naphthenic acid can intensify the corrosion [11]. At the presence of corrosive gases, the protective sulfide or oxide layer on the surface can be easily solved in the condensed compounds and bare metal is exposed to an environment having corrosive gases (especially H_2S, HCl, ammoniac, naphthenic acid and $C_6–C_7$).

Sulfide stress cracking (SSC) results from the presence of sulfur compounds such as hydrogen sulfide (H_2S) in the environment and increases its corrosivity. The presence of H_2S promotes the movement of hydrogen atoms from the metal's surface into the metal's matrix. This induces the hydrogen atoms into the structure forming hydrogen gas (H_2) at points of stress in the metal which can then lead to cracking. The following circumstances are necessary for SSC to take place:

- Susceptible material,
- Presence of tensile strength,
- Aggressive environment containing sulfur compounds, i.e. H_2S, and
- Presence of an electrolyte (moisture or water).

The equipments that are chosen to use in contact with H_2S environments should be rated for sour service (such as H_2S) with adherence to NACE MR0175 for oil and gas production environments [25]. Under the given conditions predicted by the standard, SSC can significantly be prohibited or postponed during service under H_2S condition. According to the NACE MR0175, susceptibility of the nickel-base alloys in the presence of different amount of H_2S

at the acceptable temperature is mentioned. Additionally, some parameters such as appropriate hardness under the service, time and temperature of heat treat process of each nickel-base alloy according to its application and presence of corrosive materials are explained [25].

During service, the packing made of monel 400 has been exposed to the corrosive gases containing H_2S and naphthenic Acid. H_2S can form atomic hydrogen which is absorbed on surface based on the following reactions:

$$H_2S+e^- \rightarrow HS^- + H_{ads} \tag{1}$$

$$H_{ads}+H_2S+e^- \rightarrow HS^- + H_2 \tag{2}$$

$$H_2 \rightarrow 2H_{ads} \quad \text{and} \quad H_{ads} \leftrightarrow H_{abs} \tag{3}$$

where H_{ads} is adsorbed hydrogen atoms on surface, and H_{abs} is adsorbed hydrogen atoms in monel. The equation shows that the adsorbed hydrogen on surface (H_{ads} is an exclusive source for hydrogen diffusion and accumulation in monel. If the concentration of H_2S in environment is increased, more hydrogen atom will diffuse in monel with increasing time service.

By increasing exposure time to corrosive gases, nickel and copper oxide will be formed on the surface according to following reactions:

$$Cu \leftrightarrow Cu^+ + e^- \quad \text{or} \quad 2Cu \leftrightarrow Cu^{2+} + 2e^- \tag{4}$$

$$2Cu^+ + 2HS^- \leftrightarrow Cu_2S + H_2 \tag{5}$$

$$Cu^{2+} + 2HS^- \leftrightarrow CuS_2 + H_2 \tag{6}$$

$$2Cu^+ + 2HS^- \leftrightarrow 2CuS + H_2 \tag{7}$$

$$Ni \leftrightarrow Ni^+ + e^- \tag{8}$$

$$2Ni^+ + 2HS^- \leftrightarrow 2NiS + H_2 \tag{9}$$

$$3Ni^+ + 2HS^- \leftrightarrow Ni_3S_2 + H_2 \tag{10}$$

$$3Ni^+ + 4HS^- \leftrightarrow Ni_3S_4 + 2H_2 \tag{11}$$

$$7Ni^+ + 6HS^- \leftrightarrow Ni_7S_6 + 3H_2 \tag{12}$$

As it can be seen, wide range of sulfide compounds are formed on the surface of substrate. It is expected that by the formation of sulfide layer on the surface of substrate, the sulfidation rate is reduced. However, not only a uniform sulfide layer does not form on the substrate, but also sulfidation continues during the service. Since the growth kinetics and volume of each of these sulfide compounds differ from each other, then a uniform protective sulfide layer cannot form on the surface [9], [21], [23] and [24]. Moreover, the presence of oxygen molecules, which react with H_2S and produce SO_2, intensifies corrosion phenomenon.

Due to the presence of the sulfur in the form of sulfide dioxides (SO_2), oxide compounds such as NiO and CuO can be produced according to the following reactions:

$$7Ni+2SO_2 \rightarrow 4NiO+Ni_3S_2 \tag{13}$$

$$3Ni+SO_2 \rightarrow 2NiO+NiS \tag{14}$$

$$13Ni+3SO_2 \rightarrow 6Nio+Ni_7S_6 \tag{15}$$

$$4Cu+SO_2 \rightarrow 2CuO+Cu_2S \tag{16}$$

$$3Cu+SO_2 \rightarrow 2CuO+CuS \tag{17}$$

These oxide compounds can intensify sulfidation mechanism with the formation of mixed oxide–sulfide products which are not uniform on the surface of substrate. Finally, due to presence of other corrosive gases in the crude oil, environment is worse than the situation where there is only H_2S. Also, it was suggested that the susceptibility to sulfidic corrosion may be increased by the presence of the naphthenic acid in the system. For instance, presence of naphthenic acid accelerates sulfidation of the substrate because it may solve sulfide products exposing bare metal to corrosive gases [25] and [26].

Due to presence of hydrogen atom, they diffuse in alloy and causes intergranular cracking in the sample. On the basis of the metallographical images of the failed sample (Fig. 2), cracks are present on both sides of the sample where it is exposed to H_2S and other corrosive gases, while no crack can be observed in packing before using in atmospheric tower (Fig. 3). Thus, sulfidation of packing led to SSC which failed the sample under service.

CONCLUSIONS AND RECOMMENDATIONS

In summary, presence of corrosive gases in atmospheric distilled tower, especially H_2S, HCl and naphthenic acid lead to the formation of brittle sulfide and oxide corrosion products on the surface of monel 400 packing material. Metallographical and SEM examinations revealed that microcracks appeared on the both surfaces of the packing and confirms stress corrosion cracking. XRD and EDS results showed that corrosion products mainly consist of sulfide and oxide compounds. Since this packing sample was fabricated by metal forming and machinery procedures, it might increase residual stress in final products making the sample more susceptible to this kind of corrosion.

In order to prevent or postpone this issue, following suggestions are presented [25] and [26]:

- In general, the resistance of nickel base alloys is determined by the chromium content of the material. Thus, by increasing the chromium content, resistance to sulfidation can significantly be increased.

- An appropriate thermal cycle to relieve residual stresses remained during production of the packing samples is necessary, and then an intermediate temperature heat treat process that improves hardening and strengthening of the alloy can provide the required performance in sever condition.

- In the units which have not been designed to work under the presence of naphthenic acid, use of chemical inhibitors can decline its detrimental effect.

- For severe conditions, type 317L stainless steel or other alloys with higher molybedenum (Mo) content may be required.

REFERENCES

1.   Agarwal DC, Akid R. Handbook of advanced materials-enabling new designs. Hoboken (NJ): John Wiley & Sons, Inc.; 2004.

2.   Roberge PR. Corrosion inspection and monitoring A. Ontario (Canada): John Wiley & Sons, Inc.,Royal Military College of Canada; 2007.

3.   Nasirpouri F, Alizadeh H, Hosseingholizadeh M. Failure analysis of a carbon steel screw under the service in the presence of hydrogen sulphide. Eng Fail Anal 2011;18:2316–23.

4.   Hocking MG, Sidky PS. The hot corrosion of nickel-based ternary alloys and superalloys for gas turbine applications—II. The mechanism of corrosion in SO2/O2 atmospheres. Corros Sci 1987;27:205–11.

5.   Natesan K. Oxidation–sulfidation behavior of nickel-base super alloys and M–Cr–Al coatings. Mater Sci Eng 1987;87:99–106.

6.   Lee WH. Oxidation and sulfidation of Ni3Al. Mater Chem Phys 2002;76:26–37.

7.   Quadakkers WJ, Khanna AS, Schuster H, Nickel H. Investigation of the corrosion mechanisms of nickel and nickel-based alloys in SO2-containing environments using an evolved gas analysis technique. Mater Sci Eng A 1989;120:117–22.

8.   Kutry PN, Angel RD. Isothermal sulphidation of sintered nickel-base alloys in sulphur dioxide. Corros Sci 1988;28(5):503–11.

9.   Haugsrud R, Kofstad P. On the high-temperature oxidation of Cu-rich Cu–Ni alloys. Oxid Met 1998;50:189–213.

10. Sidky PS, Hocking MG. The hot corrosion of nickel-based ternary alloys and superalloys for gas turbine applications—I. Corrosion in SO2/O2 atmospheres. Corros Sci 1987;27:183–203.

11. Kane RD, Cayard MS. A comprehensive study on naphthenic acid corrosion, corrosion 2002. USA: Nace; 2002.

12. Ali JA, Ambrose JR. The dissolution mechanism of monel 400 in Na2SO4 solutions. Corros Sci 1991;32(8):799–814.

13. Ali JA, Ambrose JR. The relationship between copper component dissolution kinetics and the corrosion behaviour of monel-400 alloy in de-aerated NaCl solutions. Corros Sci 1992;33(7):1147–59.

14. Qian W, Ishihara A, Aoyama Y, Kabe T. Sulfidation of nickel- and cobalt-promoted molybdenum–alumina catalysts using a radioisotope 35S-labeled H2S pulse tracer method. Appl Catal A 2000;196:103–10.

15. Marrone PA, Hong GT. Corrosion control methods in supercritical water oxidation and gasification processes. J Supercrit Fluids 2009;51:83–103.

16. The corrosion resistance of nickel-containing alloys in sulfuric acid and related compound. New York: The International Nickel Company, Inc.; 1983. Part 2.

17. Es-Said OS, Zakharia K, Zakharia Z, Ventura C, Pfost D, Crawford P, et al. Failure analysis of K-monel 500 (Ni ± Cu ± Al alloy) bolts. Eng Fail Anal 2000;7:323–32.

18. Rebak RB. Environmentally assisted cracking of nickel alloys – a review, second international conference on environment-induced cracking of metals (EICM-2). Banff (Alberta, Canada): Elsevier; 2004.

19. Rebak RB. Environmentally assisted cracking of commercial Ni–Cr–Mo alloys: a review, corrosion/2005 conference. Houston (TX, USA): Nace Expo; 2005.

Chapter 8

Failure Investigation of the Bucket Wheel Excavator Crawler Chain Link

Srđan M. Bošnjak[a], Dejan B. Momčilović[b],
Zoran D. Petković[a], Milorad P. Pantelić[c], and
Nebojša B. Gnjatović[a]

[a]University of Belgrade, Faculty of Mechanical Engineering, Kraljice Marije 16, 11120 Belgrade, Serbia

[b]Institute for Testing of Materials IMS, Bulevar Vojvode Mišic´a 43, 11000 Belgrade, Serbia

[c]Kolubara Metal Ltd., Diše Đur - devic´a Rusa 32, 11560 Vreoci, Serbia

ABSTRACT

The high mobility of open pit machines in heavy duty conditions provides fertile ground for the occurrence of various failures of the traveling mechanisms' vital parts such as chain links. The goal of the study presented in this paper was to diagnose the cause of the damage of the bucket wheel excavator crawler chain links. In order to identify the reasons behind chain link failures, stress state calculations were performed as well as experimental investigations which, given the nature of the failure, included visual and metallographic examinations, chemical composition analysis and tests of mechanical properties. Based on the results of the numerical–experimental analyses, it was concluded that the chain link breakdowns are caused by 'manufacturing-in' defects. The results of the presented analyses also emphasize the importance of a comprehensive quality control of chain links.

INTRODUCTION

Over the last decades earthmovers, especially bucket wheel excavators (BWEs) as continuous excavation machines and the largest structures in earth based technology, have become progressively larger and their mechanisms more efficient [1] and [2]. The natural tendency towards permanently improving the performance of the BWE, especially their capacities and mobility, has not always been adequately followed by design procedures and manufacturing technologies. This statement is substantiated by various accidents and failures of carrying structures as described and analyzed in [3], [4], [5], [6] and [7]. Similar problems occur with BW reclaimers [8] as well as various types of conveying machinery [9] and [10]. Digging drives and their vital parts [11], [12] and [13] and especially traveling mechanisms [14], [15], [16], [17], [18], [19] and [20] and belonging substructures [21] and [22], are also exposed to failure occurrence in extreme exploitation conditions. The common denominator in the failures of a wide class of machines, particularly the high-capacity BWE, is the very high financial loss, as stated in [17], [23] and [24].

The traveling mechanism of bucket wheel excavator SchRs 1760 consists of three pairs of crawlers, two of which are steerable. Constant exploitation in heavy duty conditions leads to failures of the traveling mechanism's vital substructures. Besides TWB structures [21], chain links are also exposed to different types of failures, Fig. 1. In order to detect the reason of the chain link fractures it was necessary to carry out researches which included: (a) calculation of the stress–strain state; (b) visual and metallographic inspection of crack surfaces; (c) testing of the chemical composition and mechanical properties of the chain link material.

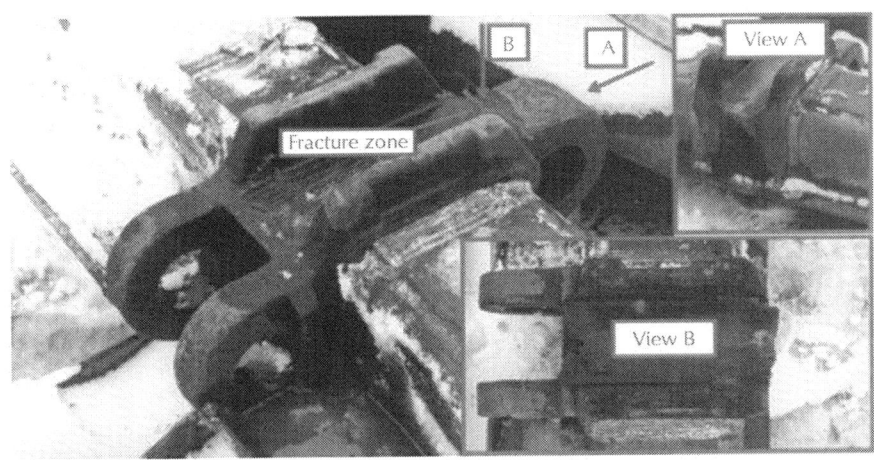

Figure 1: BWE SchRs 1760: typical chain link fracture (open pit mine "Kolubara" – Serbia).

The presented investigation results are important because of the following reasons: (a) in extremely hard working conditions, any kind of chain link disadvantages may lead to its failure [14], [15] and [16]. In this light, it is interesting to note that according to [14] the actual number of crawler segments operating in Polish brown coal mines is over 21,000 out of which about 50% with defects caused by various types of material weaknesses [14]; (b) chain link and track wheel failures cause downtimes followed by high financial losses which may exceed the direct material damage several times over [15] and [17]; (c) research literature dealing with strength problems of chain links in open pit machines is sparse. Specialized literature [1], [25] and [26], discusses solely the problems of the chain links contact strength and ground pressure distribution; (d) wide usage of crawler traveling mechanisms in various types of earthmoving and conveying machines.

FRACTURE DESCRIPTION

Two fractured segments of the chain link were examined, Fig. 2. The common feature for both samples is a fully brittle fracture originating from casting defects, like the one presented in sample 1, Fig. 2 , which shows that the crack was initiated by a shrinkage defect. The very rough surface of sample 2, detail E in Fig. 2b, indicates an increased level of microstructure inhomogeneity, causing irregular crack propagation front during the fracture. The fracture surface of sample 1 is not rough which indicates that the homogeneity of the microstructure was better compared with sample 2. An interesting feature on the fracture surface of sample 1 is the gas hole at the outer surface of the chain link, detail A in Fig. 2a, as the origin of the fracture. Secondary shrinkage, with no influence on the fracture, is visible too. On the fracture surface of sample 2, Fig. 2b (detail F), the large gas hole is apparent. Besides that, the expected signs of plastic deformation near the fracture surface, due to huge loads, were minimal, Fig. 2a (detail B). Finally, the common feature on fracture surfaces of both specimens was coarse grain.

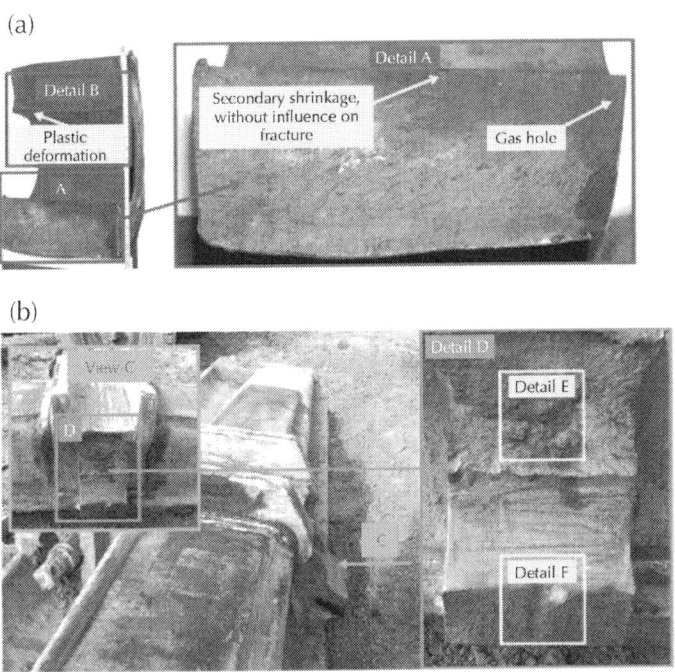

Figure 2: Typical crack surfaces of the chain link male lug: (a) sample 1; and (b) sample 2.

IDENTIFICATION OF THE STRESS STATE

Load Identification

Every chain link operates in complex interaction with other parts of the crawler travel mechanism and the ground. It is always under the influence of other chain links, while the influence of other crawler parts and the ground is present periodically, depending on its position. The load analysis, Table 1, is carried out according to the rules given in [1] and code DIN 22261-2.

Table 1: Representative Chain link loads

Nomenclature	Notation	Value (kN)
Average vertical track wheel load for maximum load on the crawler track	$R_{Z,m,max}$	384.3
Nominal chain link force	$F_{CL,nom}$	644.9
Maximum chain link force	$F_{CL,max}$	1218.8
Maximum cam force	$F_{C,max}$	1434.8

Load Cases

Loads and boundary conditions depend on the chain link position [14]. In order to investigate the stress state as a possible reason of chain link failures, the following load cases (LC) were analyzed: LC 1 – chain link is exposed to $F_{CL,nom}$; LC 2 – chain link is exposed to $F_{CL,max}$; LC 3 – chain link is exposed to $F_{CL,nom}$ and $R_{Z,m,max}$ acting above the axis of the male lug hole; LC 4 – chain link is exposed to $F_{CL,nom}$ and $R_{Z,m,max}$ acting above the root of the male lug; LC 5 – chain link is exposed to $F_{CL,nom}$ and $R_{Z,m,max}$ acting in the zone of female lug roots; LC 6 – chain link is exposed to $F_{C,max}$ during forward motion; LC 7 – chain link is exposed to $F_{C,max}$ during backward motion.

FEM Models

The FE model for stress analysis in LC 1 and LC 2 is created by discretization of the 3D chain link model. In order to truly simulate the influence of the wheel load in LC 3, LC 4 and LC 5, corresponding FE models include wheels too. In all cases, the 3D domains are discretized by the 4-node linear tetrahedron elements.

Stress States

According to the design documentation, chain links were supposed to be made from cast steel quality grade G30CrMoV6-4+QT2 (standard EN 10293). Minimum yield stress and ultimate tensile stress values are σ_{YS} = 750 MPa and σ_{UTS} = 900 MPa.

Bearing in mind that in all LC the highest stress values appear in very small zones, the domains of the chain link structure in which von Mises stresses are equal or greater than 250 MPa (1/3 of σ_{YS}) are shown in red,[1] Fig. 3. This way, the analysis of stress distribution becomes much easier.

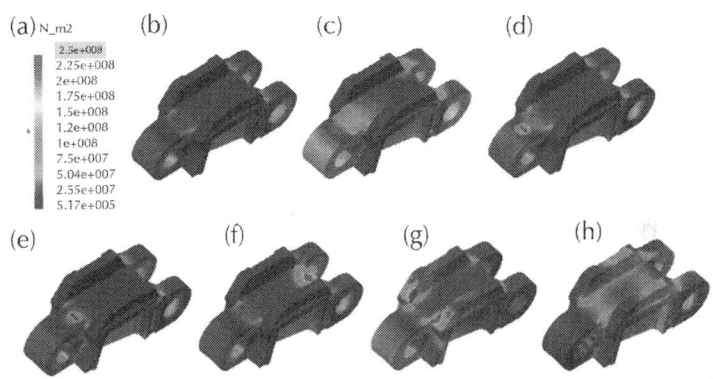

Figure 3: Von Mises stress fields: (a) stress scale; (b) LC 1, $\sigma_{vM'max}$ = 125 MPa; (c) LC 2, $\sigma_{vM'max}$ = 236 MPa; (d) LC 3, $\sigma_{vM'max}$ = 1215 MPa; (e) LC 4, $\sigma_{vM'max}$ = 1042 MPa; (f) LC 5, $\sigma_{vM'max}$ = 682 MPa; (g) LC 6, $\sigma_{vM'max}$ = 669 MPa; and (h) LC 7, $\sigma_{vM,max}$ = 407 MPa.

Maximum stress values in LC 1 and LC 2, Fig. 3b and c, are obtained in the male lug root. They are considerably lower than the minimum prescribed yield stress value, Table 2.

Table 2: Stress values in critical zones of male and female lugs

Load case	Stress values in critical zones (MPa)			
	Male lug		Female lug	
	Eye cross-section	Root	Eye cross-section	Root
1	111	125	114	92
2	210	236	219	175
3	254	201	127	154

4	106	145	124	137
5	109	141	126	120
6	164	669	109	100
7	407	261	200	238

In LC 3, LC 4 and LC 5, Fig. 3d–f, maximum stress values are obtained in the track wheel and chain link contact area. Values of the normal components of stress tensors in the vertical direction i.e. the track wheel load direction, are greater than the permissible contact stress value ($p_{per} = 1.5 \times \sigma_{YS} = 1.5 \times 750 = 1125$ MPa), but do not compromise the chain link strength. In the male lug critical zone, the values of von Mises stresses are considerably lower than the minimum prescribed yield stress value, Table 2.

The maximum stress values in critical zones are lower than the minimum prescribed yield stress value even in cases when the maximum driving force acts upon a single chain link, LC 6 and LC 7, Fig. 3g and h,Table 2.

EXPERIMENTAL PROCEDURE

Experimental investigations included: chemical analyses; tensile tests; macro and microstructural examinations; microhardness measurement. Since no fractures occurred on the female part of the chain links, specimens were sampled from the male part, according to the code EN 10293 and manufacturer recommendations.

Chemical Composition

A gravimetric analysis of the chemical composition was carried out. Chips for the analysis were taken from the cross section of specimens 1 and 2. The results of the analyses, Table 3, lead to the conclusion that the chemical composition of the chain link material does not meet the requirements of standard EN 10293 and project documentation. The content of S is much higher than required (almost 50%), while the content of Cis a bit lower.

Table 3: Chemical analysis (wt. %) of chain link material and chemical composition of G30CrMoV6-4+QT2

Material	C	Si	S	P	Mn	Ni	Cr	Mo	V	Ti	W	Al
Sample 1	0.251	0.342	0.029	0.020	0.833	0.080	1.553	0.356	0.119	–	–	0.042
Sample 2	0.246	0.414	0.028	0.016	0.902	0.055	1.391	0.386	0.123	0.012	–	0.025
G30CrMoV6-4+QT2	0.27–0.34	≤0.60	≤0.020	≤0.025	0.60–1.00	–	1.30–1.70	0.30–0.50	0.05–0.15	–	–	–

Tensile Testing

Tensile tests were carried out on round specimens (8 mm in diameter) in accordance with standard EN 10002-1. It can be taken as conclusive (Table 4) that for sample 1 specimens the yield stress values are considerably lower (from 9.1% to 15.7%) than the minimal values required by standard EN 10293, while tensile strength values are lower for 12.8–13.7% than the minimal values required by the mentioned standard. Strength characteristics of specimens taken from sample 2 meet the standard requirements but the elongation is lower for 20.8–40.7%.

Table 4: Tension test results and tensile properties of G30CrMoV6-4+QT2

Sample	Specimen	σ_{YS} (MPa)	σ_{UTS} (MPa)	Elongation A5 (%)	Contraction (%)
1	1	682	781	11.75	37.98
	2	646	776	12.50	36.00
	3	632	784	13.50	29.85
2	1	915	1005	6.30	14.40
	2	933	1034	9.50	27.70
	3	920	1010	7.50	21.30
G30CrMoV6-4+QT2		min. 750	900–1100	min. 12	–

Macrostructure and Microstructure Examinations

A bainite – ferrite microstructure with numerous non-metallic inclusions is detected on sample 1, Fig. 4a. A similar microstructure with an appearance of porosity was revealed on sample 2, Fig. 4b. On the MTG sample taken from the specimen used for tensile testing, numerous non-metallic inclusions and cracks can be observed, Fig. 5.

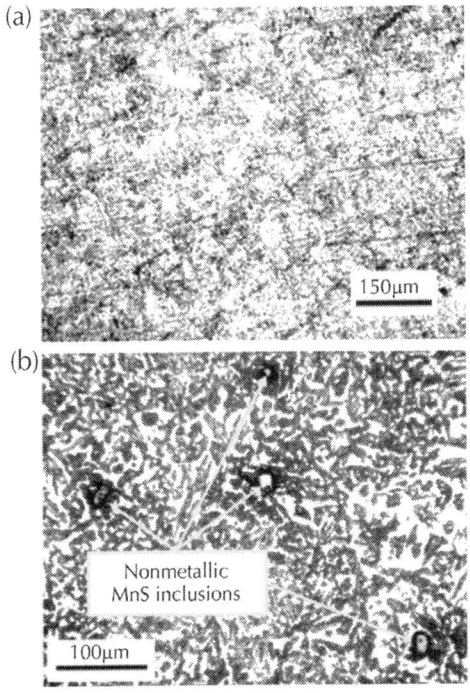

Figure 4: Microstructures: (a) sample 1; and (b) sample 2 (etched with 3% nital).

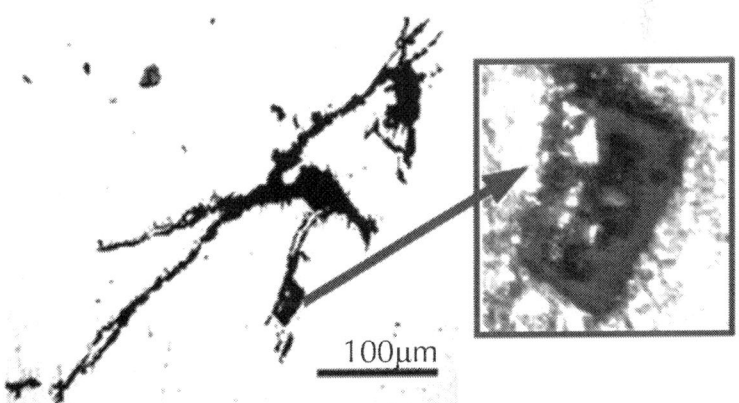

Figure 5: MTG sample (framed detail: large inclusion MnS).

Microhardness Measurement

The determination of hardness was carried out by microhardness (HV1) testing of specimens for metalographic examinations. Testing was performed as per the SRPS ISO 6507-1/2011 standard by means of a semiautomatic optical instrument HAUSER 249A. The results of the measurements show that the hardness of sample 1 is relatively uniform along the cross-section (305–313 HV1), while on sample 2 there are two locations with considerably lower hardness values (214 and 219 HV1) as opposed to the dominant part of the cross-section where hardness values differ from 305 to 331 HV1.

DISCUSSION

Based on the results of stress analyses presented in Section 3.4, it is conclusive that:

- In all LC, stress values in the male lug root are greater than the corresponding stress values in the female lug roots from 5.8% in LC 4 to 569.0% in LC 6, Table 2, Fig. 6 and Fig. 7. Stress values in the female lug eye cross-section are greater than corresponding stress values in the male lug eye cross-section from 2.6% in LC 1 to 14.5% in LC 4. The largest differences are obtained for relatively low stress values, lower than 130 MPa, so they are not significant. In LC 3, LC 6 and LC 7 the situation is vice versa: stress values in the male lug eye cross-section are greater for 100.0%, 50.5% and 103.5%, respectively. Furthermore, the greatest differences are obtained in LC 3 and LC 7 in which the highest stress values occur, 254 MPa and 407 MPa, respectively.

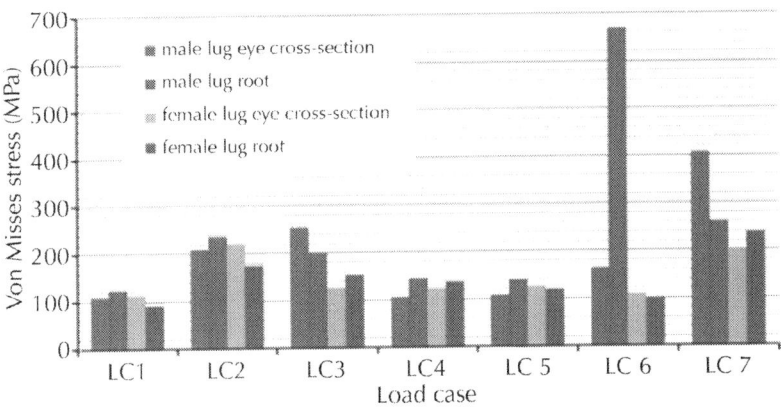

Figure 6: Maximum von Mises stress values in critical zones.

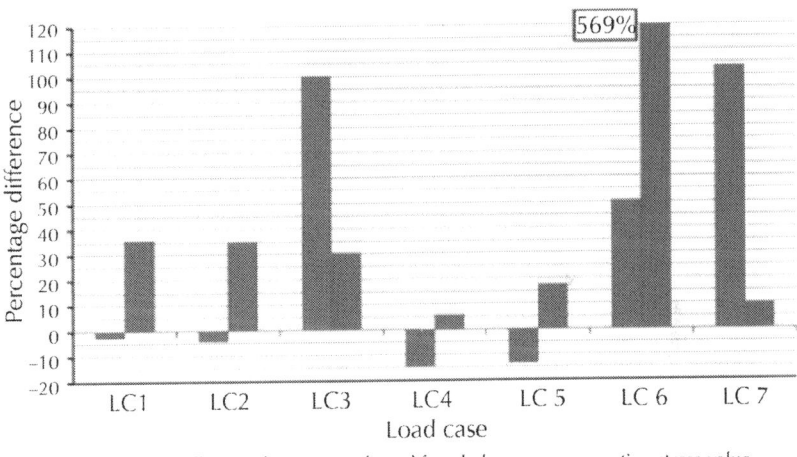

Figure 7: Percentage difference of stress values in critical zones.

- Very high values of von Mises stresses in LC 3, LC 4 and LC 5 do not jeopardize the chain link integrity because the contact stresses are dominant. Even in the worst case (LC 3) the zone of high stress values is relatively small (depth d ≈ 3.7 mm, width w ≈ 17.1 mm), Fig. 8. In addition to that, stress state identification was done under the assumption that the average vertical track wheel load for maximum load on the crawler track is acting on the chain link, which appears relatively rarely.

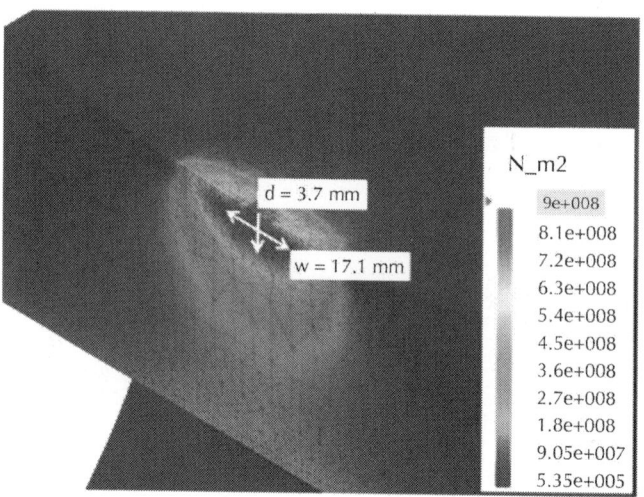

Figure 8: Distribution of von Mises stresses in the eye cross-section in LC 3 (the zone of stress values higher than $_{UTS}$ = 900 MPa is colored in red). (For interpretation of the references to color in this figure legend, the reader is referred to the web version of this article.)

The increased sulphur content of both samples, compared with specifications, Table 3, means that the chain link will have decreased toughness, particularly under impact conditions.

The significant decrease of elongation compared with specified values, Table 4, confirms the presumptions based on the results of the chemical composition. The obtained low values of elongation and contraction mean that sample 2 had very low resistance to crack initiation and crack propagation. In the case of casting defects, Fig. 2, the first overload lead to instantaneous crack initiation and crack propagation i.e. an immediate fracture.

CONCLUSIONS

Chain links are designed in full accordance with all operational demands. Although contact stresses are higher than the permissible values, stress states determined under the assumption that there are no material-in faults, cannot be the cause of drastic volumetric chain link destruction. Chain link fractures are localized in zones of high calculated stress values, which represent indirect validation of the used calculating method. The fact that chain link fractures appear only in the zones of male lugs can be explained by: (a) higher stress values in the male lug root in all load cases; (b) higher stress values in the male lug eye cross-section in load cases with highest stress fields; (c) considerably larger cross-section dimensions as opposed to the female lug, consequently leading to a higher probability of material-in defect occurrences, which were detected on fractured surfaces.

Based on all facts presented, it is conclusive that the considered chain links failures are predominantly caused by the so-called 'manufacturing-in' defects [24] and [27]. The results of the presented analyses also emphasize the importance of a comprehensive quality control of chain links. Any aberration from the check list that should cover the NDT and mechanical test, could mask the presence of casting defects and deficiency of mechanical properties and subsequently lead to premature failure.

ACKNOWLEDGMENTS

This work is a contribution to the Ministry of Education, Science and Technological Development of Serbia funded Project TR 35006.

REFERENCES

1. Durst W, Vogt W. Bucket wheel excavator. Clausthal-Zellerfeld: Trans Tech Publications; 1989.

2. Singh S. The state of the art in automation of earthmoving. J Aerosp Eng 1997;10:179–88.

3. Bošnjak S, Zrnic´ N, Simonovic´ A, Momcˇilovic´ D. Failure analysis of the end eye connection of the bucket wheel excavator portal tie-rod support. Eng Fail Anal 2009;16:740–50.

4. Rusin´ski E, Czmochowski J, Iluk A, Kowalczyk M. An analysis of the causes of a BWE counterweight boom support fracture. Eng Fail Anal 2010;17:179–91.

5. Bošnjak S, Petkovic´ Z, Zrnic´ N, Simic´ G, Simonovic´ A. Cracks, repair and reconstruction of bucket wheel excavator slewing platform. Eng Fail Anal 2009;16:1631–42.

6. Bošnjak S, Pantelic´ M, Zrnic´ N, Gnjatovic´ N, Ðor - devic´ M. Failure analysis and reconstruction design of the slewing platform mantle of the bucket wheel excavator O&K SchRs 630. Eng Fail Anal 2011;18:658–69.

7. Jovancˇic´ P, Ignjatovic´ D, Tanasijevic´ M, Maneski T. Load-bearing steel structure diagnostics on bucket wheel excavator, for the purpose of failure prevention. Eng Fail Anal 2011;18:1203–11.

8. Araujo LS, de Almeida LH, Batista EM, Landesmann A. Failure of a bucket-wheel stacker reclaimer: metallographic and structural analyses. J Fail Anal and Prev 2012;12:402–7.

9. Maneski T, Jovancˇic´ P, Ignjatovic´ D, Miloševic´ -Mitic´ V, Maneski M. Condition and behaviour diagnostics of drive groups on belt conveyors. Eng Fail Anal 2012;22:28–37.

10. Bošnjak S, Zrnic´ N. Dynamics, failures, redesigning and environmentally friendly technologies in surface mining systems. Arch Civ Mech Eng 2012;12:348–59.

11. Rusin´ ski E, Harnatkiewicz P, Kowalczyk M, Moczko P. Examination of the causes of a bucket wheel fracture in a bucket wheel excavator. Eng Fail Anal 2010;17:1300–12.

12. Savkovic´ M, Gašic´ M, Arsic´ M, Petrovic´ R. Analysis of the axle fracture of the bucket wheel excavator. Eng Fail Anal 2011;18:433–41.

13. Arsic' M, Bošnjak S, Zrnic' N, Sedmak A, Gnjatovic' N. Bucket wheel failure caused by residual stresses in welded joints. Eng Fail Anal 2011;18:700–12.

14. Dudek D, Frydman S, Huss W, Pe˛kalski G. The L35GSM cast steel – possibilities of structure and properties shaping at the example of crawler links. Arch Civ Mech Eng 2011;11:19–32.

15. Bošnjak S, Arsic' M, Zrnic' N, Odanovic' Z, Ðor - devic' M. Failure analysis of the stacker crawler chain link. Procedia Eng 2011;10:2244–9.

16. Zhi-wei Y, Xiao-lei X, Xin M. Failure investigation on the cracked crawler pad link. Eng Fail Anal 2010;17:1102–9.

17. Arsic' M, Bošnjak S, Odanovic' Z, Dunjic' M, Simonovic' A. Analysis of the spreader track wheels premature damages. Eng Fail Anal 2012;20:118–36.

18. Rusin' ski E, Czmochowski J, Moczko P. Half-shaft undercarriage systems – designing and operating problems. JAMME 2009;33:62–9.

19. Clegg RE. Failure of planetary pinions in earth moving equipment – a failure analysis approach. Eng Fail Anal 2000;7:35–41.

20. Savkovic' M, Gašic' M, Petrovic' D, Zdravkovic' N, Pljakic' R. Analysis of the drive shaft fracture of the bucket wheel excavator. Eng Fail Anal 2012;20:105–17.

21. Bošnjak S, Petkovic' Z, Zrnic' N, Pantelic' M, Obradovic' A. Failure analysis and redesign of the bucket wheel excavator two-wheel bogie. Eng Fail Anal 2010;17:473–85.

22. Bošnjak S, Simonovic' A, Petkovic' Z, Zrnic' N. Comparative analysis of strength for variant structural solutions of lower structure for bucket wheel excavator KRUPP C-700S. J Appl Eng Sci (former title: Istraz̆ivanja i projektovanja za privredu) 2006;4:19–28.

23. Wintle JB, Pargeter RJ. Technical failure investigation of welded structures (or how to get the most out of failures). Eng Fail Anal 2005;12:1027–37.

24. Bošnjak S, Arsic´ M, Zrnic´ N, Rakin M, Pantelic´ M. Bucket wheel excavator: integrity assessment of the bucket wheel boom tie – rod welded joint. Eng Fail Anal 2011;18:212–22.

25. Der RasperL. Sshaufelradbagger als gewinnungsgerat. Clausthal-Zellerfeld: Trans Tech Publications; 1975.

26. Pajer G, Pfeifer M, Kurth F. Tagebaugroßgeräte und universalbagger. Berlin: VEB Verlag Technik; 1971.

27. Gagg CR. Failure of components and products by 'engineered-in' defects: case studies. Eng Fail Anal 2005;12:1000–26.

Metabolic Engineering for Production of Biorenew-able Fuels and Chemicals: Contributions of Synthetic Biology

Laura R. Jarboe[1], Xueli Zhang[2], Xuan Wang[2], Jonathan C. Moore[2], K. T. Shanmugam[2], and Lonnie O. Ingram[2]

[1]Department of Chemical and Biological Engineering, Iowa State University, 3051 Sweeney Hall, Ames, IA 50011, USA

[2]Department of Microbiology and Cell Science, University of Florida, P.O. Box 110700, Gainesville, FL 32611, USA

ABSTRACT

Production of fuels and chemicals through microbial fermentation of plant material is a desirable alternative to petrochemical-based production. Fermentative production of biorenewable fuels and chemicals requires the engineering of biocatalysts that can quickly and efficiently convert sugars to target products at a cost that is competitive with existing petrochemical-based processes. It is also important that biocatalysts be robust to extreme fermentation conditions, biomass-derived inhibitors, and their target products. Traditional metabolic engineering has made great advances in this area, but synthetic biology has contributed and will continue to contribute to this field, particularly with next-generation biofuels. This work reviews the use of metabolic engineering and synthetic biology in biocatalyst engineering for biorenewable fuels and chemicals production, such as ethanol, butanol, acetate, lactate, succinate, alanine, and xylitol. We also examine the existing challenges in this area and discuss strategies for improving biocatalyst tolerance to chemical inhibitors.

INTRODUCTION

Human society has always depended on biomass-derived carbon and energy for nutrition and survival. In recent history, we have also become dependent on petroleum-derived carbon and energy for commodity chemicals and fuels. However, the nonrenewable nature of petroleum stands in stark contrast to the renewable carbon and energy present in biomass, where biomass is essentially a temporary storage unit for atmospheric carbon and sunlight-derived energy. Thus there is increasing demand to develop and implement strategies for production of commodity chemicals and fuels from biomass instead of petroleum. Specifically, in this work we are interested in the microbial fermentation of biomass-derived sugars to commodity fuels and chemicals.

In order for a fermentation process to compete with existing petroleum-based processes, the target chemical must be produced at a high yield, titer and productivity. Sometimes there are additional constraints on the fermentation process, such as the presence of potent inhibitors in biomass hydrolysate or the need to operate at an extreme pH or temperature [1]. These goals can be difficult to attain with naturally-occurring microbes. Therefore, microorganisms with these desired traits often must be developed, either by modification of existing microbes or by the de novo design of new microbes. While significant progress has been made towards de novo design [2, 3], this work focuses on the modification of existing microbes.

Humanity has long relied on microbial biocatalysts for production of fermented food and beverages and eukaryotic biocatalysts for food and textiles. We have slowly modified these biocatalysts by selecting for desirable traits without understanding the underlying biological mechanisms. But upon elucidation of the biological code and the development of recombinant DNA technology, we now have the tools to do more than just select for observable traits—we are now able to rationally modify and design metabolic pathways, proteins, and even whole organisms.

Much of this rational modification has been in the form of Metabolic Engineering. Metabolic Engineering was defined in 1991 [4, 5] and here we use the definition of "the directed improvement of production, formation, or cellular properties through the modification of specific biochemical reactions or the introduction of new ones with the use of recombinant DNA technology" [6]. While Metabolic Engineering has enabled extraordinary advances in the production of commodity chemicals and fuels from biomass, some of which are discussed in this work, we have now reached the point where biological functions that do not exist in nature are desired. Synthetic biology aims to develop and provide these non-natural biological functions.

For many years, the term Synthetic Biology was used to describe concepts that would be classified today as Metabolic Engineering [7]. However in the last 10 years, terms such as "unnatural organic molecules" [7], "unnatural chemical systems [8], "novel behaviors" [9], "artificial, biology-inspired systems" [10], and "functions that do not exist in nature" [11] have been used to describe Synthetic Biology. For the purpose of this review, we will apply the Synthetic Biology definition of "the design and construction of new biological components, such as enzymes, genetic circuits, and cells, or the redesign of existing biological systems" [12].

Synthetic biology has application to many fields, including cell-free synthesis [13], tissue and plant engineering [14] and drug discovery [15], but here we are interested in the modification of microbes for the bio renewable production of commodity chemicals and fuels. Other recent reviews have also dealt with this topic [16–18].

Synthetic biology for the production of a target compound can be expressed as a sequence of the following events, each of which will be discussed in more detail and demonstrated below. (1) Design the metabolic pathways and phenotypic properties of the desired system. What are the desired substrates and products? What are the expected environmental stressors? (2). Choose an appropriate host organism (chassis) based on the following criteria. Which organisms display at least some of the desired properties? How well characterized and annotated are these organisms? Are there molecular biology tools for modification of this chassis? (3). Formulate an implementation approach. What modifications are necessary to achieve the pathways and properties identified in step (1)? Do metabolic pathways need to be added, removed, or tuned? Does the desired pathway or phenotype exist in nature, or does it need to be designed de novo? (4). Optimize the redesigned system and assess the system properties relative to the ideal. Can the chassis be improved further?

Even a simple biocatalyst, such as the laboratory workhorse Escherichia coli, is a complex system of an estimated 4603 genes, 2077 reactions, and 1039 unique metabolites [19, 20], and while the steps outlined above are relatively straightforward, it is still difficult to quickly and reliably engineer a biocatalyst to perform desired behaviors [21]. Systems biology, the standardization of biological systems, and metabolic evolution are all vital to the compensation for this disconnect between the expected and actual biocatalyst behaviors. Through a combination of these powerful techniques, biocatalysts have been redesigned for the production of an astounding array of commodity fuels and chemicals, both natural and unnatural (Figure 1 and Table 1). Here we discuss successful examples involving the production of commodity fuels and chemicals, with a focus on D- and L-lactate, L-alanine, succinate, ethanol, and butanol.

Table 1: Summary of engineered E. coli biocatalysts for production of renewable fuels and chemicals in our laboratory

Product	Fermentation condition[1]	Titer (g/L)	Yield(g/g)	Productivity (g/L/h)	Reference
Redesign through modification of existing pathways					
D-lactate	Anaerobic, batch	118	0.98	2.88	[22]
Acetate	Aerobic, fed-batch	53	0.50	1.38	[23]
Succinate	Anaerobic, batch	83	0.98	0.90	[24]
Redesign through introduction of foreign pathways					
Ethanol	Anaerobic, batch	43	0.48	2.00	[25]
L-lactate	Anaerobic, batch	116	0.98	2.29	[22]
Xylitol	Aerobic, fed-batch	38	1.40	0.81	[26]
L-alanine	Anaerobic, batch	114	0.95	2.38	[27]

[1]All fermentations were done in mineral salts medium with glucose, except for the ethanol fermentations which used xylose.

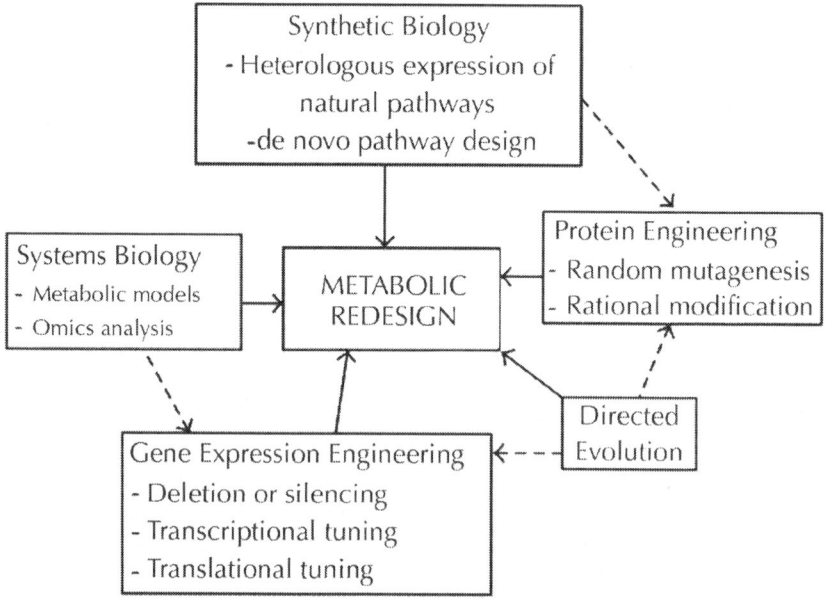

Figure 1: Overview of tools for metabolic redesign.

METHODS AND TOOLS FOR BIO-CATALYST REDESIGN

Chassis

A robust and stable chassis enables efficient and economical production of fuels and chemicals at an industrial level. Since we are specifically interested in biocatalysts that can utilize biomass, a desirable chassis has the following characteristics: (1) growth in mineral salts medium with inexpensive carbon sources, (2) utilization of hexose and pentose sugars, so that all the sugar components in lignocellulosic biomass can be converted to the desired product, (3) high metabolic rate, essential for high rate of productivity,

(4) simple fermentation process to reduce the manipulation cost and minimize failure risks in large-scale production, (5) robust organism (high temperature and low pH where possible) to reduce the requirement for external cellulase during cellulose degradation, as well as to reduce the required amount of base addition, (6) ease of genetic manipulation and genetic stability, (7) resistance to inhibitors produced during the biomass pretreatment process, and (8) tolerance to high substrate and product concentrations in order to obtain high titers of target compound.

Enteric bacteria, especially E. coli, have many of the above mentioned physiological characteristics and are, thus, an excellent chassis for synthetic biology. Most of the examples discussed here use E. coli, but other important microbial model systems have been redesigned, including Clostridium acetobutylicum [28], Corynebacterium glutamicum [29], Saccharomyces cerevisiae [30], and Aspergillus Niger [31]. E. coli has been used as a model organism since the beginning of genetic engineering [32]. While K-12 strain MG1655 (ATCC# 47076) is one of the most commonly used E. coli strains [33], there are other lineages, such as B (ATCC# 11303), C (ATCC# 8739), and W (ATCC# 9637), that are also generally regarded as safe since they are unable to colonize the human gut [34]. Although K-12 is the most characterized and widely used strain, E. coli W (ATCC# 9637) and C (ATCC# 8739) have proven to be better chassis for synthesizing fuels and chemicals. For example, K-12-derived strains were unable to completely ferment 10% (w/v) glucose in either complex or mineral salts medium [1, 35], while derivatives of strains W or C can completely ferment more than 10% (w/v) of glucose with higher cell growth and sugar utilization rates than K-12. Additionally, E. coli W strains have the native ability to ferment sucrose [1, 36].

Foreign genes may be unstable in host cells due to recombination facilitated by mobile DNA elements, and thus the mobile DNA elements in E. coli K-12 strain have been deleted [37]. This minimal genome construction strategy is an excellent approach to improve this chassis for the production of fuels and chemicals.

Systems Biology Tools

Genome-Scale Models and In Silico Simulation

Given the rational basis of metabolic engineering and synthetic biology, models and simulations are critical predictive and tools. Genome sequencing and automatic annotation tools have enabled construction of genome-scale metabolic models of nearly 20 microorganisms [38]. These constraint-based models and in silico-simulations can be used to predict metabolic flux redistribution after genetic manipulation, or to predict other cellular functions, such as substrate preference, outcomes of adaptive evolution and shifts in expression profiles [39]. They can also aid in pathway design to obtain desired phenotypes [40–42]. For example, the E. coli iJE660a GSM model was used to successfully simulate single- and multiple-gene knockouts to improve lycopene production [42]. The computational framework, Optknock, was developed to identify gene deletion targets for system optimization [41], and simulation results for gene deletions for succinate, lactate, and 1,3-propanediol production were in agreement with experimental data. Another simulation program, OptStrain, was developed to guide metabolic pathway modification for target compound production, through both the addition of heterologous metabolic reactions and deletion of native reactions [40]. However, most of the current models only have stoichiometric information, while kinetic and regulatory effects are not included [38,39]. Integration of kinetic and regulatory information will improve the accuracy and predictive power of these models.

High-Throughput Omics Analysis

High-throughput omics analysis, such as transcriptome, proteome, metabolome, and fluxome [43–45], aids in characterization of cellular function on multiple levels, and therefore provide a "debugging" capability for system optimization [12, 45]. Genetic manipulations can disturb the metabolic balance or impair cell growth due to depletion of important precursors [46, 47], accumulation of toxic intermediates [48], or redox imbalance [1]. For example, high NADH levels in E. coli reengineered for ethanol production inhibited citrate synthase activity, thereby limiting cell growth by lowering production of the critical metabolite 2-ketoglutarate [49]. Metabolome and fluxome analysis can quickly identify the limiting metabolites or altered metabolic flux distribution, providing the basis for problem solving [45, 50]. For example, metabolite measurements of Aspergillus terreus were implemented in the rational metabolic redesign for increased production of lovastatin [45, 50]. Changes of mRNA and protein profiles can be identified by transcriptome and proteome analysis, providing gene targets for further engineering [46, 47]. The work of Choi et al. demonstrate this concept: transcriptome analysis of E. coli producing the human insulin-like growth factor I fusion protein aided in selection for targets for gene deletion. The resulting redesigned strain showed a greater than 2-fold increase in product titer and volumetric productivity [46, 47]. Additionally, comparative genome sequence analysis facilitates identification of mutated genes or regulators during evolution, and these mutations can be used to redesign the systems for better synthetic capability. For example, in an effort described as "genome-based strain reconstruction", evolved strains of Corneybacterium glutamicum selected for L-lysine production were compared to the parental strain, and mutations were found that were proposed as beneficial to L-lysine production. Three of these mutations were introduced into the parent strain and enabled production of up to 3.0 g/L/hr L-lysine [51].

Genetic Manipulation Tools

Gene Deletion

Gene deletion can redistribute carbon flux toward the target product by deleting genes critical to competing metabolic pathways and, thus, is widely used in metabolic redesign strategies. Homologous recombination is the most frequently used strategy for gene-deletion (Figure 2). Historically, plasmids containing a selectable marker flanked by DNA fragments homologous to the target gene and either temperature sensitive or conditional replicons were needed for efficient gene deletion in bacteria [52] (Figure 2(a)). In contrast, genes can be directly disrupted in yeast by linear PCR fragments with short flanking DNA fragments homologous to chromosomal DNA. Linear DNA is not as easy to transform into E. coli because of the intracellular exonuclease system and low recombination efficiency. Gene deletion systems based on bacteriophage λ Red recombinase facilitate chromosomal gene deletion using a linear PCR fragment [53]. In this method, the chromosomal gene is replaced by the selectable marker flanked by two FRT (FLP recognition target) fragments (Figure 2(b)) and then the marker can be removed by the FLP recombinase [54]. However, this method leaves a 68bp-FRT scar on the chromosome after each excision [52], reducing further gene deletion efficiency. Repeated use of this FRT/FLP system for specific gene deletions has the potential to generate large unintended chromosomal deletions.

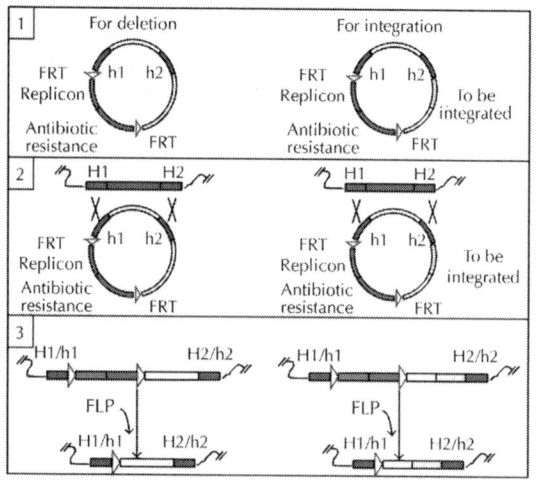

(a)

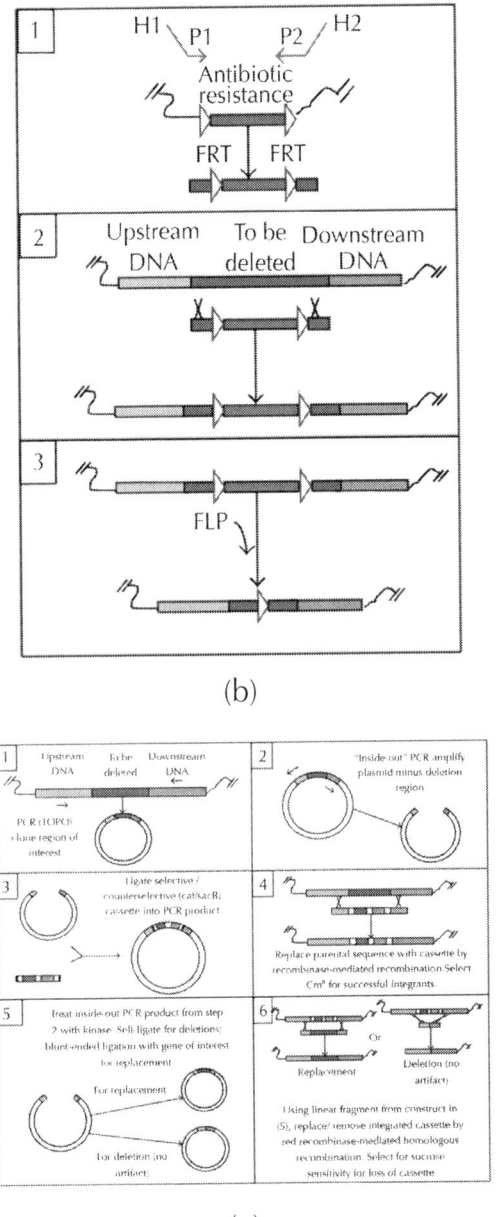

(b)

(c)

Figure 2: Comparison of three-gene deletion methods in E. coli. These methods can also be used in other enteric bacteria. The first and third methods can also be used for gene integration into the chromosome and promoter replacement for tuning gene expression. 2(a) plasmid-based method. Step 1 is construction of the deletion plasmid containing DNA fragments homologous to the target gene (h1 and h2), a selectable marker, and either a temperature sensitive or conditional replicon. Step 2 is double-crossover recombination; the plasmid cannot replicate in the host strain, and antibiotic-resistant colonies are selected. In step 3, the FRT, replicon, and antibiotic resistance marker are removed by FLP. 2(b) Linear DNA-based method. Step 1 is construction of the linear DNA fragment by PCR (H1-P1 and H2-P2 as primers). H1 and H2 refer to short DNA fragments homologous to target gene. Step 2 is replacement of the target gene with the antibiotic resistance gene through crossover recombination with the help of Red recombinase. Step is removal of FRT and antibiotic marker by FLP. 2(c) two-stage recombination-based method developed in our lab. Steps, 2, 3, and 5 describe construction of the plasmids and linear DNA fragments for the two-stage recombinations. Step1, 2, 3 and 5 describes the first recombination step, in which the cat, sacB cassette is inserted into the target gene. Step 4 is the second recombination step, in which the cat, sacB cassette is removed by selection on sucrose.

To facilitate sequential gene deletions, our lab has developed a two-stage recombination strategy (Figure 2(c)), using the sensitivity of E. coli to sucrose when Bacillus subtilis levansucrase (sacB) is expressed [24, 27, 55]. Gene deletions created by this method do not leave foreign DNA, antibiotic resistance markers, or scar sequences at the site of deletion. In the first recombination, part of the target gene is replaced by a DNA cassette containing a chloramphenicol resistance gene (cat) and levansucrase gene (sacB). In the second recombination, the cat, sacB cassette is removed by selection for resistance to sucrose. Cells containing the sacBgene accumulate levan during incubation with sucrose and are killed [55]. Surviving recombinants are highly enriched for loss of the cat, sacB cassette [24, 27].

Gene Expression Tuning

Like gene deletions, plasmid-based expression systems are ubiquitous to metabolic redesign. However, plasmid-based systems have several disadvantages. (1) Plasmid maintenance is a metabolic burden on the host cell, especially for high-copy number plasmids

[56]. Note that high copy numbers are not essential, considering that most central metabolic enzymes are encoded by a single gene; (2) plasmid-based expression is dependent on plasmid stability, with only few natural unit-copy plasmids having the desired stability [12]; (3) only low-copy number plasmids have replication that is timed with the cell cycle, and thus maintaining a consistent copy number in all cells is challenging [12]; (4) metabolic redesign can require construction of a complex heterologous pathway, and thus several genes, encoded in large pieces of DNA, need to be incorporated. Most commercial plasmids have difficulties carrying large DNA fragments.

Chromosomal integration of the target genes followed by fine-tuning their expression could eliminate these plasmid-associated problems. The abovementioned two-step recombination strategy for gene deletion can also be used for gene integration or promoter replacement (Figure 2).

Gene expression in prokaryotes is mainly controlled at the transcriptional level, and therefore the promoter is the most tunable element. While inducible promoters, such as lac and ara, have been traditionally used to modulate gene expression, large-scale inducer use is cost prohibitive for production of fuels and bulk chemicals. However, several strategies have been developed to construct constitutive promoter libraries for fine-tuning gene expression. Some methods rely on the use of natural promoters. For example, Zymomonas mobilis genomic DNA was used to construct a promoter library for screening optimal expression of Erwinia chrysanthemi endoglucanase genes (celY and celZ) in Klebsiella oxytoca P2 in order to improve ethanol production from cellulose [57]. Other methods rely on random modification of existing promoters, such as the randomization of the spacer sequences between the consensus sequences [58], or mutagenesis of a constitutive promoter [59]. This promoter modification method was used to assess the impact of phosphoenolpyruvate carboxylase levels on cell yield and deoxy-xylulose-P synthase levels on lycopene production, and the optimal expression levels of these genes were identified for maximal desired phenotype [59].

These synthetic promoter libraries could also be integrated into the chromosome directly, which could facilitate expression modulation of chromosomal genes [60, 61].

The fine-tuning methods described above rely on the selection of the best natural promoter or random alteration of existing promoters. One of the goals of synthetic biology is construction of standard parts, and posttranscriptional processes, such as transcriptional termination, mRNA degradation, and translation initiation, have been engineered with this goal in mind. Examples include construction of a synthetic library of 5' secondary structures to successfully manipulate mRNA stability [62], and modulation of the ribosome binding site (RBS) as well as Shine-Dalgarno (SD) and AU-rich sequences to tune gene expression at the translation initiation process [60, 63]. Riboregulators were also developed to tune gene expression via RNA-RNA interactions [64]. A final method of fine-tuning gene expression is codon optimization, which can improve translation of foreign genes [65]. These optimized gene sequences often do not exist in nature and must be generated using DNA synthesis techniques.

In many cases, more than one gene needs to be introduced into the chassis and expression of these genes needs to be coordinated to attain desired biocatalyst performance. One such method is modulation of the expression of each individual gene via its own promoter. However, it is difficult to predict the appropriate expression level of each gene. Another option is to combine multiple genes into a synthetic operon with a single promoter, and fine-tune expression of each gene through posttranscriptional processes [12] with tunable control elements (such as mRNA secondary structure, RNase cleavage sites, ribosome binding sites, and sequestering sequences) at intergenic regions. Libraries of tunable intergenic regions (TIGRs) were generated and screened to tune expression of several genes in an operon [48]. This method was used to coordinate expression of three genes in an operon that encodes a heterologous mevalonate biosynthetic pathway, improving mevalonate production by 7-fold [48]. Another method to control expression of more than one gene is to engineer global transcription machinery by random mutagenesis of transcription factors [66, 67]. This method was shown to efficiently improve tolerance to toxic compounds and production of metabolites, and to alter phenotypes [66, 67].

Protein Engineering

Natural proteins may not meet the required criteria for specific and efficient system performance, and thus alteration for a specific application may be needed. Directed evolution of proteins offers a way to rapidly optimize enzymes, even in the absence of structural or mechanistic information [68]. For directed evolution, a protein library is usually generated by random mutagenesis [68], recombination of a target gene [69], or a family of related genes [70] and then the library is analyzed by high-throughput screening. This method has been used to successfully increase enzyme activity [71, 72], increase protein solubility and expression, invert enantioselectivity, and increase stability and activity in unusual environments [68]. For example, a mutation library of the gene-encoding geranylgeranyl diphosphate synthase of Archaeoglobus fulgidus was generated to screen for mutants with higher activity, enabling lycopene production in E. coli. Screening of more than 2,000 variants identified eight with increased activity; one of which increased lycopene production by 100% [71]. Of particular relevance to the field of synthetic biology is the creation of novel enzymatic activity through protein engineering [73, 74]. For example, the unnatural isomerization of α-alanine to β-alanine was attained by evolving a lysine 2,3-aminomutase to expand its substrate specificity to include α-alanine [73].

Rational design is another powerful tool to increase protein properties, especially with the aid of computational analysis [75, 76]. Based on knowledge of protein structure and function, one can predict which amino acid(s) to change in order to obtain the desired function. In the redesign of Lactobacillus brevis for the production of secondary alcohols, it was desired to change the cofactor preference of the R-specific alcohol dehydrogenase from

NADPH to NADH. A structure-based computational model was used to identify potentially beneficial amino acid substitutions and one of these changes increased NADH-dependent activity four-fold [77].

While these examples demonstrate the power of rational enzyme (re-)design, this approach requires detailed information about the protein structure and mechanism, while random mutagenesis does not. Recent advances have combined directed evolution and rational design in a so-called "semi-rational" approach to successfully improve enzyme activity when only limited information is available [78, 79]. When the mutagenesis is limited to specific residues, as chosen from existing structural or functional knowledge, these "smart" libraries are more likely to yield positive results [79]. For example, the catalytic activity of pyranose-2-oxidase was improved by mutagenesis of the known active site [80].

While the 20 natural amino acids supply enzymes with a wide range of possible activity, this range can be expanded even further by the use of unnatural amino acids (UAAs). There are more than 40 UAAs available at this time and they have been used to probe protein function, photocage critical residues, and alter metalloprotein properties [81, 82]. While this technology is still in the developmental stage, at least one study has shown an improvement in enzyme activity following insertion of UAAs. Site 124 of E. coli's nitroreductase was replaced with a variety of natural and unnatural amino acids and certain UAA variants had a greater than 2-fold increase in activity over the best natural amino acid variant [83]. This biomimetic approach has been expanded to other metabolites, such as carbohydrates [84] and lipids [85].

Evolution

As described above, a robust biocatalyst with high yield, titer, and productivity is critical for a fermentation process to compete with petrochemical-based production. Current models and simulation tools provide a framework given the constraints of known protein functions. But the many reactions and enzymes that remain uncharacterized cannot be included in this theoretical analysis. Therefore rational design methods often result in a biocatalyst that performs poorly relative to the model. Metabolic evolution provides a complementary approach to improve biocatalyst productivity and robustness, dependent upon the design of an appropriate selection pressure. Where feasible, synthesis of the target compound can be coupled to the production of ATP, redox balance, or key metabolites that are essential for growth, and selection for improvements in growth during metabolic evolution (serial transfers) can be used to coselect for higher rates or titers of target compounds (Figure 3). Both redox balance and net ATP production in such a synthetic system are requisites for successful evolution.

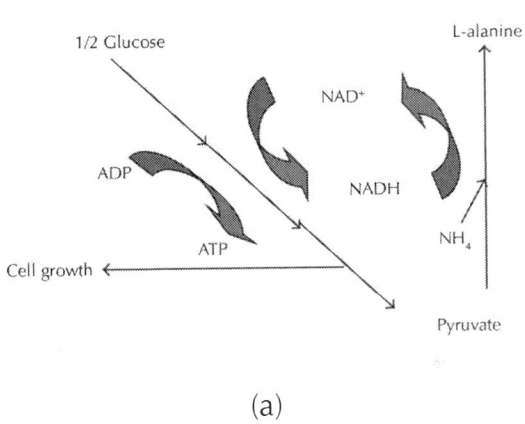

(a)

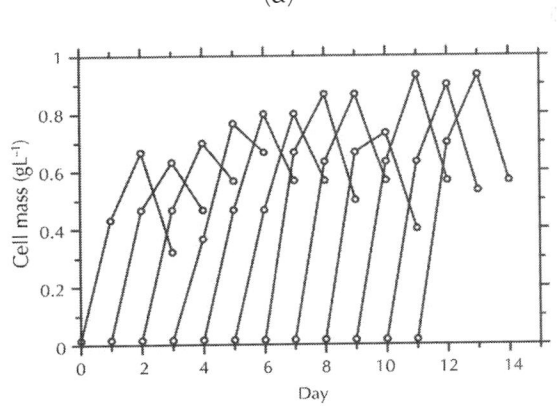

(b)

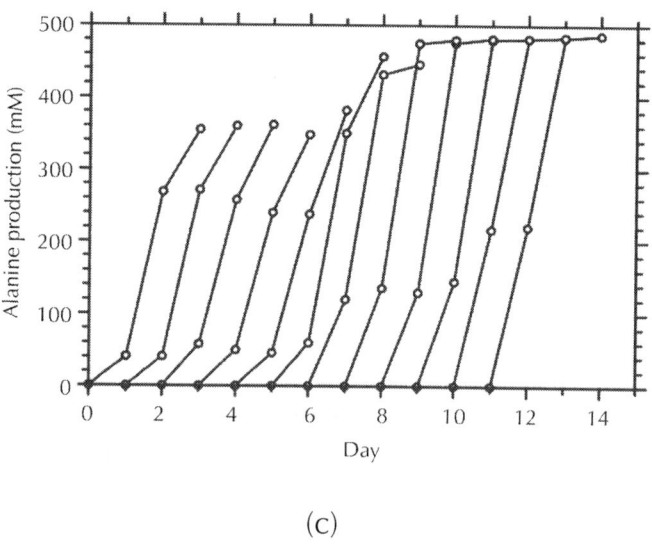

(c)

Figure 3: Metabolic evolution for improving L-alanine production in E. coli [27]. 3(a) Redesigned metabolic pathway for L-alanine production: ATP production and cell growth is coupled to NADH oxidation and L-alanine production. 3(b) Directed evolution improves cell growth. Parental strain XZ112 reaches a maximum cell mass of 0.7 gL^{-1} after 48 hours of fermentation; evolved strain XZ113 attains 0.7 gL^{-1} after 24 hours and a maximum of 0.9 gL^{-1} after 48 hours; 3(c) metabolic evolution to improve cell growth also improves alanine production. Parental strain XZ112 produces 355 mM alanine after 72 hours of fermentation; evolved strain XZ113 produces 484 mM in 48 hours.

We have used this metabolic evolution strategy to optimize biocatalysts redesigned for production of several fermentation products [1], including ethanol, D-lactate, L-lactate, L-alanine (Figure 3), and succinate, as described in more detail below. A frequently-used design scheme is to couple synthesis of the target product to growth by inactivating competing NADH-consuming pathways. Thus, the only way for cells to regenerate NAD$^+$ for glycolysis is to produce the target compound. Increased cell growth, supported by higher ATP production rate during glycolysis, is coupled with higher NADH oxidization rate, and thus tightly coupled with synthesis of target product. This evolution strategy has been shown to increase productivity by up to two orders of magnitude. Computational frameworks based on genome-scale metabolic models have been used to construct biocatalysts that couple biomass formation with chemical production [40, 41], and therefore provide a basis for selective pressure for high productivity. For example, Optknock identified gene deletion targets for the construction of lactate-producing E. coli, and then directed evolution improved production capability [86]. Although rational design of metabolic pathways based on current metabolic models is a common method for maximizing yield of the target compound, this method is not always the best strategy, due to our limited understanding of the complicated metabolic network and dynamic kinetics of each reaction. Metabolic evolution provides an excellent alternative method for strain improvement, through which reactions that are not currently predictable would be selected to improve biocatalyst performance [87]. As our knowledge of biocatalyst behavior and metabolism improves, predictive models will become even more powerful.

REDESIGN THROUGH MODIFICATION OF EXISTING PATHWAYS

In this section, we highlight projects that have redesigned a chassis to produce target compounds at high yield and titer without the introduction of foreign pathways. In the next section, we describe biocatalyst redesigns which used foreign or nonnatural pathways.

Succinate

Succinate, a four-carbon dicarboxylic acid, is currently used as a specialty chemical in food, agricultural, and pharmaceutical industries [88] but can also serve as a starting point for the synthesis of commodity chemicals used in plastics and solvents, with a potential global market of $15 billion [89]. Succinate is primarily produced from petroleum and there is considerable interest in the fermentative production of succinate from sugars [89].

Several rumen bacteria can produce succinate from sugars with a high yield and productivity [90–92], but require complex nutrients. Alternatively, native strains of E. coli ferment glucose effectively in simple mineral salts medium but produce succinate only as a minor product [93]. Therefore E. coli strain C (ATCC 8739) was redesigned for succinate production at high yield, titer, and productivity [94].

The initial redesign strategy focused on inactivation of competitive pathways, specifically deletion of lactate dehydrogenase (ldhA), alcohol/aldehyde dehydrogenase (adhE), and acetate kinase (ackA). However, the resulting strain grew poorly in mineral salts medium under anaerobic condition and accumulated only trace amounts of succinate. Because NADH oxidization is coupled to succinate synthesis in this strain, metabolic evolution was used to improve both the cell growth and succinate production. After inactivation of pyruvate formate-lyase and methylglyoxal synthase to eliminate formate and lactate production, the final strain, KJ073, produced near 670 mM succinate (80 g/L) in mineral salts medium with a high yield (1.2 mol/mol glucose) and high productivity (0.82 g/L/h) [94]. Inactivation of threonine decarboxylase (tdcD), 2-ketobutyrate formate-lyase (tdcE), and aspartate aminotransferase (aspC) further increased succinate yield (1.5 mol/mol glucose), titer (700 mM), and productivity (0.9 g/L/h) [24].

Despite its power in improving biocatalyst performance, metabolic evolution has the undesirable property of being a black box; evolved strains show the desired biocatalyst properties, but the metabolic evolution process does not improve our understanding of the biocatalyst. Therefore, reverse engineering of evolved strains can help us identify key mutations that can then be rationally applied to other biocatalysts. Reverse engineering of the succinate-producing strain revealed two significant changes in cellular metabolism that increased energy efficiency [87]. The first change is that PEP carboxykinase (pck), which normally functions in gluconeogenesis during the oxidative metabolism of organic acids [90, 95, 96], became the major carboxylation pathway for succinate production. High-level expression of PCK dominated CO_2 fixation and increased ATP yield (1 ATP per oxaloacetate produced). The second change is that the native phosphoenolpyruvate- (PEP-) dependent phosphotransferase system for glucose uptake was inactivated and replaced by an alternative glucose uptake pathway: GalP permease (galP) and glucokinase (glk). These changes increased the pool of PEP available for maintaining redox balance, as well as increasing energy efficiency by eliminating the need to produce additional PEP from pyruvate, a reaction that requires two ATP equivalents [97].

While rational design based on current metabolic understanding is a key component of metabolic engineering and synthetic biology, our limited understanding of the complicated metabolic network and dynamic kinetics of each reaction can lead to failure of predictive models. In this example, metabolic evolution was demonstrated as an excellent alternative method for strain improvement, through which currently unpredictable reactions would be selected to expand cellular metabolic capability [87]. By understanding the mutations that enabled desirable performance of the succinate-producing strain, we have more options available for the redesign of future systems. To demonstrate this, E. coli was again redesigned based on the findings from the evolved strain [98]. This time, the design strategy shifted from inactivating competitive fermentation pathways to recruiting energy conserving pathways for efficient succinate production (Figure4 (e)). After increasing pck gene expression and inactivating the native glucose PTS system, the native E. colimetabolic system was converted to an efficient succinate synthetic system, equivalent to the native pathway of succinate-producing rumen bacteria [98].

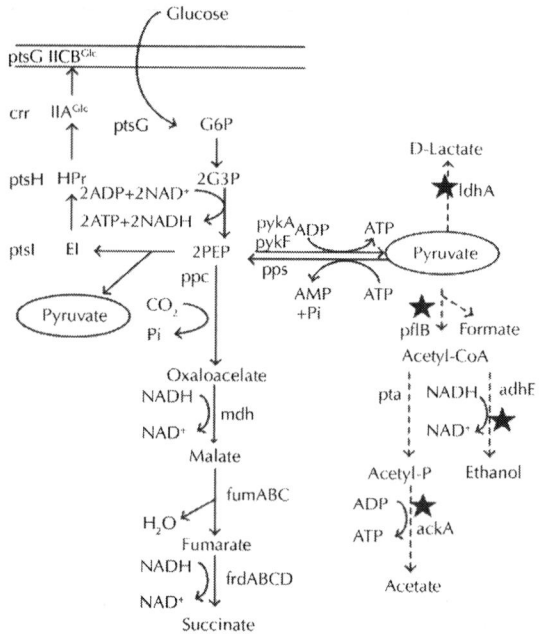

(a)

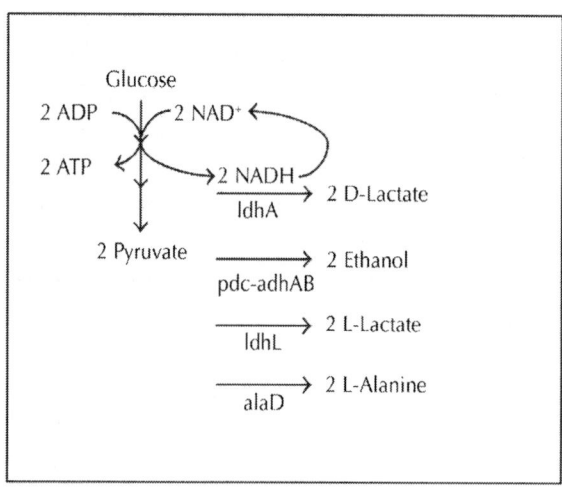

(b)

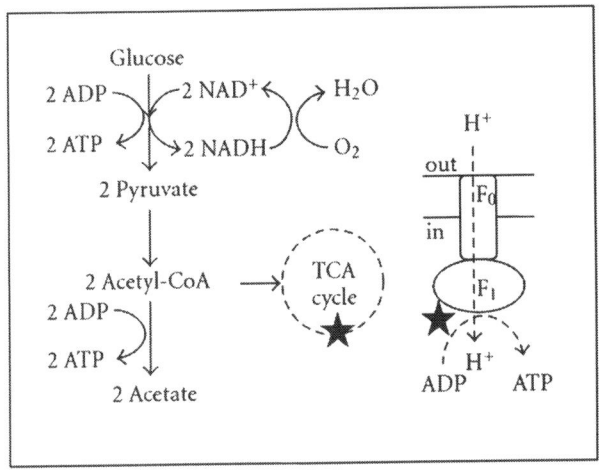

(c)

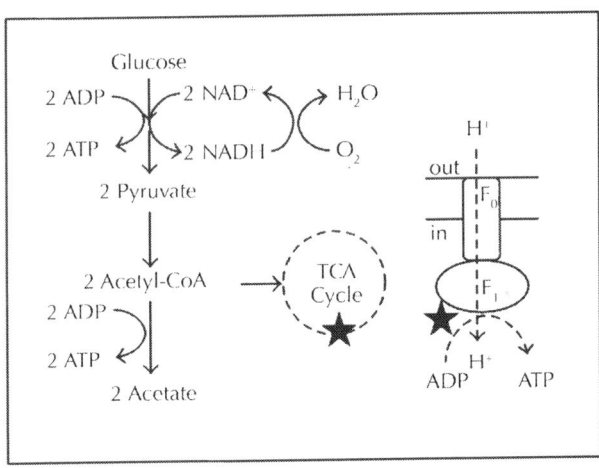

(d)

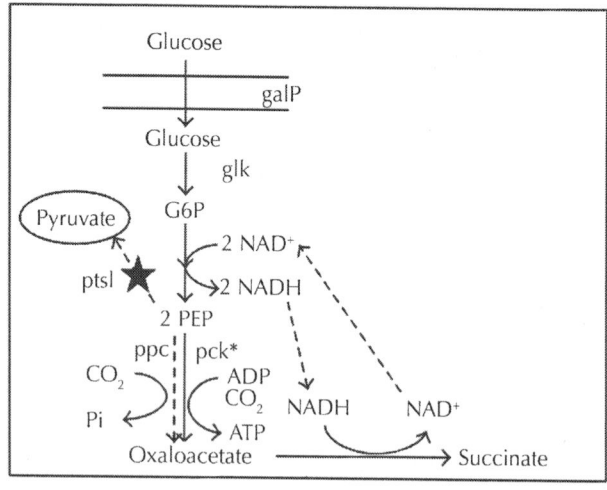

(e)

Figure 4: Synthetic pathways of E. coli for production of fuels and chemicals in our lab: 4(a) Native metabolic pathways of glucose fermentation in E. coli; 4(b) synthetic pathways for production of D-lactate, ethanol, L-lactate and L-alanine; 4(c) synthetic pathways for production of pyruvate and acetate; 4(d) synthetic pathway for production of xylitol, 4(e) synthetic pathway for production of succinate. ★ indicate gene deletion. Genes and enzymes: ackA, acetate kinase; adhAB, alcohol dehydrogenase (Z. mobilis); adhE, alcohol/aldehyde dehydrogenase; alaD, L-alanine dehydrogenase (G. stearothermophilus);crr, glucose-specific phosphotransferase enzyme IIA component; frd, fumarate reductase;fum, fumarase; galP, galactose-proton symporter (glucose permease); glk, glucokinase;ldhA, D-lactate dehydrogenase; ldhL, L-lactate dehydrogenase (P. acidilactici); mdh, malate dehydrogenase; pdc, pyruvate decarboxylase; pflB, pyruvate formate-lyase; ppc, phosphoenolpyruvate carboxylase; pta, phosphate acetyltransferase; ptsG, PTS system glucose-specific EIICB component; ptsH, phosphocarrier protein HPr; ptsI, phosphoenolpyruvate-protein phosphotransferase (Phosphotransferase system, enzyme I); pyk, pyruvate kinase; xrd, xylose reductase (C. boidinii); xylB, xylulokinase. Metabolites: G6P, glucose-6-phosphate; G3P, glycerol-3-phosphate; PEP, phosphoenol pyruvate; X5P, D-xylulose-5-phosphate.

D-Lactate

D-lactate is widely used as a specialty chemical in the food and pharmaceutical industry. It can also be combined with L-lactate for the production of polylactic acid (PLA), an increasingly popular biorenewable and biodegradable plastic [99, 100] whose commercial success obviously depends on the production cost. Although glucose is the current substrate for fermentative production of lactate, it is desirable to produce this commodity chemical from lignocellulosic feedstock, which contains a mixture of sugars. Some lactic acid bacteria have the desirable native ability to produce large amount of D-lactate under low pH condition, where the low pH reduces the process cost [101, 102]. However, these lactic acid bacteria require complex nutrients, and many of them lack the ability to ferment pentose sugars. The lactic acid bacteria that do ferment pentose sugars unfortunately produce a mixture of lactate and acetate and, thus, are not a good chassis for commercial production of D-lactate. While E. coli can ferment many sugars effectively in a simple mineral salts medium, inherent D-lactate productivity is low and other undesirable metabolites are also produced [103]. Therefore, the E. coli metabolic system was redesigned to attain the desired properties of high yield and productivity of D-lactate.

E. coli strain W3110 was used as the chassis for D-lactate production with a redesign strategy that focused on inactivation of competitive fermentation pathways [104]. After deleting the genes encoding fumarate reductase (frdABCD), alcohol/aldehyde dehydrogenase (adhE), pyruvate formate lyase (pflB), and acetate kinase (ackA), the resulting strain, SZ63, can only oxidize NADH via D-lactate synthesis (Figure 4(b)). Although this strain could completely utilize 5% (w/v) glucose in a mineral salts medium with a yield near theoretical maximum (96%), the volumetric D-lactate productivity of 0.42 g/L/h was relatively low compared with lactic acid bacteria [35]. In addition, this strain can neither utilize sucrose nor completely utilize 10% (w/v) sugar [35]. Therefore, an E. coli W derivative strain was chosen as chassis for more robust D-lactate production [35, 105]. After redesigning central metabolism so that D-lactate production was the sole means of oxidizing NADH, metabolic evolution was used to further improve cell growth and D-lactate productivity. The resulting strain, SZ194, efficiently consumed 12% (w/v) glucose in mineral salts medium and produced 110 g/L D-lactate [105] with a volumetric productivity of 2.14 g/L/h, a 5-fold increase over the W3110 derivative. The biocatalyst was further optimized by deleting methylglyoxal synthase gene (mgsA) to eliminate L-lactate production, and by metabolic evolution to increase yield and productivity. The final D-lactate producing strain, TG114, could convert 12% (w/v) glucose to 118 g/L D-lactate with an excellent yield (98%) and productivity (2.88 g/L/h) [22].

Acetate

Acetate is a commodity chemical with 2001 worldwide production estimated at 6.8 million metric tons [23]. Biological production of acetate accounts for only 10% of world production, mainly in the form of vinegar, with the remainder of production through petrochemical routes [106–108]. Biological production of commodity chemicals has historically focused on anaerobic production of reduced products, since substrate loss as cell mass and CO_2 is minimal and product yields are high. Contrastingly, acetate is an oxidized chemical, and traditional biological production involves a complex two-stage process: fermentation of sugars to ethanol by Saccharomyces, followed by aerobic oxidation of ethanol to acetate by Acetobacter [106–108]. To enable microbial production of redox-neutral or oxidized products at high yield, the biocatalyst metabolism needs to be redesigned to combine attributes of both fermentative and oxidative metabolisms.

Redesign of E. coli W3110 metabolism for acetate production focused on three major pathways: fermentative metabolism, oxidative metabolism, and energy supply (Figure 4(c)) [23]. The competitive fermentation pathways (pflB, ldhA, frd, adhE) were inactivated to prevent the consumption of common precursor pyruvate, and the oxidative tricarboxylic acid (TCA) cycle was interrupted to reduce the carbon loss as CO_2. Finally, oxidative phosphorylation was disrupted (atpFH) to reduce ATP production while maintaining the ability to oxidize NADH by the electron transport system, thus increasing the glycolytic flux for more ATP production through substrate-level phosphorylation. Although rationally designed, the resulting strain, TC32, had an undesirable auxotrophic requirement for succinate during growth in glucose-minimal medium. Evolution was used to eliminate this auxotrophy and the final strain, TC36, produced 878 mM acetate (53 g/L) in mineral salts medium with 75% of the maximal theoretical yield. Although this is a lower titer than acetate produced from ethanol oxidation by Acetobacter, TC36 has a two-fold higher production rate, requires only mineral salts medium, and can metabolize a wide range of carbon sources in a simple one-step process [23].

Others

Butanol is an excellent alternative transportation fuel with several advantages compared to ethanol, including higher-energy content, lower volatility, less hydroscopicity, and less corrosivity [109]. Redesign of E. coli for butanol production is discussed below. C. acetobutylicum ATCC 824 naturally produces butanol and was redesigned to increase butanol production and decrease coproduct accumulation. Metabolic engineering-type modifications, such as overexpression of the acetone formation pathway to increase formation of butanol precursor butyryl-CoA, inactivation of the transcriptional repressor SolR, and overexpression of alcohol/aldehyde dehydrogenase all increased butanol production [110–112]. In an excellent example of synthetic biology-type applications, expression of the butyrate kinase gene was fine-tuned by a rationally designed antisense RNA to increase butanol production [113].

1, 2-propanediol (1, 2-PD) is a major commodity chemical currently derived from propylene. E. coli naturally produces low amounts of 1,2-PD, and therefore its metabolism was redesigned to produce 1,2-PD at high yield and titer from glucose this was achieved by inactivation of competing pathways (lactate dehydrogenase and glyoxalase I), and overexpression of essential genes of 1,2-PD synthetic pathway (methylglyoxal synthase, glycerol dehydrogenase, and 1,2-PD oxidoreductase) [114]. Evolution was also used in combination with rational design for increased 1, 2-PD production [115]. L-valine, an essential hydrophobic and branched-chain amino acid, is used in cosmetics, pharmaceuticals, and animal feed additives [116]. E. coli was redesigned for L-valine production at high yield and titer from glucose through a combination of traditional metabolic engineering and synthetic biology. Traditional metabolic engineering was used to inactivate competing pathways and overexpress acetohydroxy acid synthase I (ilvBN), part of the valine biosynthesis pathway. Unfortunately, the E. coli chassis has regulatory elements that tightly control L-valine biosynthesis, making production of valine at high yield and titer difficult. Feedback inhibition was eliminated by rational site-directed mutagenesis of acetohydroxy acid synthase III. In an excellent demonstration of the gene expression tuning techniques discussed above, transcriptional attenuation of valine biosynthesis genes ilvGMEDA was eliminated by replacing the attenuator leader region with the constitutivetac promoter. Transcriptome analysis and in silico simulation guided selection of additional target genes for amplification and deletion, and the final biocatalyst produced 0.378 g L-valine per g glucose, giving a titer of 7.55 g/L valine from 20%(w/v) glucose [116]. A similar strategy was also used for L-threonine production [117].

REDESIGN THROUGH INTRODUCTION OF FOREIGN OR NONNATURAL PATHWAYS

Foreign Pathways

Ethanol

Ethanol is a renewable transportation fuel. Replacement of gasoline with ethanol would significantly reduce US import oil dependency, increase the national security, and reduce environmental pollution [118]. However, only 9 billion gallons of ethanol were produced in 2008, and all were from corn-based production. Lignocellulose is generally regarded as an excellent source of sugars for conversion into fuel ethanol. It is, thus, desirable to design or obtain biocatalysts that can utilize all the sugar components in lignocellulose and convert them to ethanol with high yield and productivity in mineral salts medium. Native S. cerevisiae and Z. mobilisstrains can efficiently convert glucose to ethanol, but cannot utilize pentose sugars. In contrast, E. coli strains can utilize all the sugar components of lignocelluloses but ethanol is only a minor fermentation product, with mixed acids accumulating as the major fermentation product [103]. While recent advances have been made engineering the native E. coli metabolic pathways for ethanol production [119], the most successful example used a foreign metabolic pathway to enable ethanol production from E. coli strain W (ATCC# 9637) [1].

Redesign for ethanol production was decoupled to three parts: construction of a metabolic pathway for production of ethanol as the major fermentation product, elimination of competitive NADH oxidization pathways, and disruption of side-product formation. The Z. mobilis homoethanol pathway (pyruvate decarboxylase and alcohol dehydrogenase) was introduced as a foreign pathway, enabling redox-balanced production of ethanol at high yield [120] (Figure 4(b)). Then fumarate reductase (frd) was disrupted to increase ethanol yield. The resulting strain, KO11, produced ethanol at a yield of 95% in a complex medium [121]. This strain was developed at the dawn of metabolic engineering and has been used to produce ethanol from a variety of lignocellulosic materials, as reviewed in [1].

Although the ethanol production rate of KO11 was as high as yeast, the ethanol tolerance and performance in minimal medium did not meet the desired standards. Therefore strain SZ110, a derivative of KO11 modified for lactate production in mineral salts media [35], was redesigned for ethanol production [122]. As with the design of KO11, redesign of SZ110 was decoupled to construction of an ethanol synthetic pathway, elimination of competitive NADH oxidization pathways, and blockage of side-product formation. However, this redesign strategy also included the acceleration of mixed sugar co-utilization. The lactate producing pathway was disrupted and the Z. mobilis homoethanol pathway was integrated into the chromosome by random insertion to select for optimal expression. The Pseudomonas putida short-chain esterase (estZ) [123] was introduced to decrease ethyl acetate levels in the fermentation broth and decrease the downstream purification cost. In addition, methylglyoxal synthase (mgsA) was inactivated, resulting in co-metabolism of glucose and xylose, and accelerated the metabolism of a 5-sugar mixture (mannose, glucose, arabinose, xylose, and galactose) to ethanol [25]. After using evolution to increase cell growth and production, the final strain, LY168, could concurrently metabolize a complex combination of the five principal sugars present in lignocellulosic biomass with a high yield and productivity in mineral salts medium [25].

L-Lactate

As described above, L-lactate is the major component of the biodegradable plastic PLA. Although many lactic acid bacteria produce L-lactate with high yield and productivity [124], they usually require complex nutrients.E. coli does not have a native pathway for L-lactate production, and therefore introduction of a foreign pathway was necessary.

The strategy for redesigning E. coli W3110 for L-lactate production was to eliminate competitive NADH oxidization pathways and then construct the desired L-lactate synthetic pathway (Figure 4(b)) [125]. The L-lactate production pathway, L-lactate dehydrogenase (ldhL) from Pediococcus acidilactici, was used and its coding region and terminator were integrated into the E. coli chromosome at the ldhA site, so that ldhL could be expressed under the native ldhA promoter. In addition, since the ldhL gene contains a weak ribosomal-binding region, this region was rationally replaced with ldhA's RBS [125]. Following a period of metabolic evolution, the resulting strain, SZ85, synthesized 45 g/L L-lactate in a mineral salts medium with yield near theoretical maximum (94%). However, this strain was a K-12 derivative and displayed the same problems seen with the K12-based D-lactate-producing strain described above, meaning that it was unable to completely ferment high sugar concentrations and had a low productivity (0.65 g/L/h). Therefore, the same design strategy was implemented in an E. coli W (ATCC# 9637) derivative. After further deleting mgsA gene to improve chiral purity and using metabolic evolution to improve cell growth and productivity, the final L-lactate-producing strain, TG108, could convert 12% glucose to 116 g/L L-lactate with an excellent yield (98%) and productivity (2.29 g/L/h) [22].

Xylitol

The pentahydroxy sugar alcohol xylitol is commonly used to replace sucrose in food and as a natural, non-nutritive sweetener that inhibits dental caries [126]. Xylitol can also be used as a building block for synthesizing new polymers [127]. Current xylitol commercial production involves hydrogenation of hemicellulose-derived xylose with an active metal catalyst [127]. Biological-based processes have also recently been developed, but although high xylitol titer was achieved by some yeast, the process requires complex medium with numerous expensive vitamin supplements [128]. While E. coli does not have the native capability to synthesize xylitol, a redesign strategy for strain W3110 was proposed involving a foreign metabolic pathway [26]. In the proposed redesign, glucose would support cell growth and provide reducing equivalents, while xylose would be used as substrate for xylitol synthesis (Figure 4(d)). The design strategy consisted of three major components: enabling co-utilization of glucose and xylose, separation of xylose metabolism from central metabolism, and construction of a xylitol production pathway (Figure 4(d)). In order to enable co-utilization of glucose and xylose, glucose-mediated repression of xylose metabolism was eliminated by replacing the native crp gene with a cAMP-independent mutant (CRP*). Xylose metabolism was separated from central metabolism by deleting the xylulokinase (xylB) gene, preventing the loss of xylose carbon to central metabolism. Finally, xylose reductase and xylitol dehydrogenase from several microorganisms were tested for xylitol synthetic capability, and the NADPH-dependent xylose reductase from C. boidinii (CbXR) was found to support optimal xylitol production. The final strain, PC09 (CbXR), could produce 250 mM (38 g/L) xylitol in mineral salts medium. The yield was 1.7 mol xylitol per mol glucose consumed, which was improved to 4.7 mol/mol by using resting cells. It was proposed that xylitol production could be further improved by increasing supply of reducing equivalents [129].

L-Alanine

L-alanine can be used with other L-amino acids as a pre- and postoperative nutrition therapy in pharmaceutical and veterinary applications [130]. It is also used as a food additive because of its sweet taste. The annual worldwide production of L-alanine is around 500 tons [131], and this market is currently limited by production costs. The current commercial production process converts aspartate to alanine via aspartate decarboxylase, where aspartate is produced from fumarate by aspartate ammonia-lyase catalysis [27]. An efficient fermentative process with a renewable feedstock such as glucose offers the potential to reduce L-alanine cost and facilitate a broad expansion of the alanine market into other products.

SZ194, a derivative of E. coli W (ATCC# 9637) that was previously engineered for D-lactate production, was used as the chassis for L-alanine production [27] (Figure 4(b)). Alanine production in the native strain uses glutamate- and NADPH-dependent glutamate-pyruvate aminotransferase. It is preferable to produce L-alanine directly from pyruvate and ammonia using an NADH-dependent enzyme, and therefore L-alanine dehydrogenase (alaD) of Geobacillus stearothermophilus was employed. The native ribosome binding site, coding region, and terminator of alaD gene were integrated into the E. coli chromosome at the ldhA site, so that expression of alaD could be controlled by the native promoter of ldhA, a promoter that has worked well for production of D- and L-lactate, as described above. Further redesign focused on elimination of trace amounts of lactate and increasing the L-alanine chiral purity by deleting mgsA and the major alanine racemase gene (dadX). Metabolic evolution increased the final titer and productivity by 15- and 30-fold, respectively (Figure 3). The latest L-alanine producing strain, XZ132, converted 12% glucose to 114 g/L L-alanine with a 95% yield and the excellent volumetric productivity of 2.38 g/L/h [27].

Combining Multiple Foreign Pathways in a Single Chassis

Although the work described above relied on the introduction of a single foreign pathway, there are other excellent examples that employ pathways from more than one organism in a single host.

E. coli was redesigned for 1,3-propanediol production using S. cerevisiae pathway to convert glucose to glycerol and a K. pneumonia pathway to convert glycerol to 1,3-propanediol [132]. E. coli was also redesigned for isopropanol production by combining acetyl CoA acetyltransferase (thl) and acetoacetate decarboxylase (adc) from C. acetobutylicum with the second alcohol dehydrogenase (adh) from C. beijerinckii and E. coli's own acetoacetyl-CoA transferase (atoAD) [133]. Artemisinic acid, a precursor of antimalarial drug artemisin, was produced by E. coli following the combination of a mevalonate pathway from S. cerevisiae and E. coli, amorphadiene synthase, and a novel cytochrome P450 monooxygenase (CYP71AV1) from Artemisia annua[12, 134].

S. cerevisiae was redesigned for flavanone production by combining Arabidopsis thaliana cinnamate 4-hydroxylase (C4H), Petroselinum crispum 4-coumaroyl: CoA-ligase (4CL), and Petunia chalcone synthase (CHS), Petunia chalcone isomerase (CHI) [135]. A similar synthetic system producing hydroxylated flavonols was also constructed in E. coli with additional amplification of C. roseus P450 flavonoid , -hydroxylase (F35H) fused with P450 reductase, Malus domestica flavanone 3, 5-hydroxylase (FHT), and Arabidopsis thalianaflavonol synthase (FLS) [136]. The flavonoid production was significantly increased through further redesigning of the central metabolic system of E. coli to increase precursor (Malonyl-CoA) supply [137].

Modification of Natural Pathways for Production of Unnatural Compounds

One of the goals of synthetic biology is to design or construct new genetic circuits. In the examples given thus far, existing biological parts have been reassembled to engineer a biocatalyst that efficiently produces a product that already exists in nature. However, metabolic pathways can also be constructed to produce unnatural compounds.

As discussed above, directed evolution of proteins can modify their activity such that new substrates are recognized or new products are formed [138]. For example, novel carotenoid compounds were generated by evolution of two key carotenoid synthetic enzymes, phytoene desaturase, and lycopene cyclase [139]. Additionally, combinatorial biosynthesis, which combines genes from different organisms into a heterologous host, can also generate new products [140]. For example, four previously unknown carotenoids were produced by combinatorial biosynthesis in E. coli [141].

De Novo Pathway Design

In order to broaden the available biosynthesis space, it is essential to go beyond the natural pathways and design pathways de novo [142]. Although this exciting design strategy still has many challenges, several successful examples have been reported.

For example, a synthetic pathway for 3-hydroxypropionic acid (3-HP) production was designed involving the unnatural isomerization of α-alanine to β-alanine, as mentioned above. In this example the researchers used directed evolution to expand the substrate specificity of lysine 2,3-aminomutase to include α-alanine [73]. The resulting β-alanine can then be converted to 3-HP through existing metabolic pathways.

Unnatural pathways for higher alcohol production in E. coli were designed by combining the native amino acid synthetic pathways with a 2-keto acid decarboxylase from Lactococcus lactis and alcohol dehydrogenase fromS. cerevisiae [143]. The 2-keto acid intermediates in amino acid biosynthesis pathways were redirected from amino acid production to alcohol production, enabling production of 3-methyl-1-pentanol. This pathway was then expanded for production of unnatural alcohols by rational redesign of two enzymes, with the resulting biocatalysts having the ability to synthesize various unnatural alcohols ranging in length from five to eight carbons [144].

Engineering Tolerance to Inhibitory Compounds

As our repertoire of biologically-produced compounds increases, tolerance to high product titers becomes more important. Biofuels, such as ethanol and butanol, can inhibit biocatalyst growth, and therefore the tolerance of the biocatalyst needs to be improved [145–147]. As described above, our goal is to use lignocellulosic biomass as a substrate for production of commodity fuels and chemicals. Unfortunately, the processes used to convert biomass to soluble sugars also produce a mixture of minor products, such as furfural and acetic acid that inhibit biocatalyst metabolism [148]. Although most of these inhibitors could be removed by detoxification [149], this additional process would increase operational cost. It is, thus, desirable to obtain microorganisms that are tolerant to these inhibitors and can directly ferment hemicellulose hydrolysate.

One approach to increasing tolerance is to understand the mechanism of inhibition. Transcriptome analysis has been used to probe the response to ethanol [145, 150], furfural [151], and butanol [147]. Another approach is to use directed evolution, as highlighted by the following example. Ethanologenic E. coli strain LY180 (a derivative of LY168 with restored lactose utilization and integration of an endoglucanase, and cellobiose utilization) was used as the chassis to select for furfural resistance through evolution [148]. The evolved strain, EMFR9, had significantly increased furfural resistance. Reverse engineering efforts, including transcriptome analysis, attributed furfural resistance to the silencing expression of several oxidoreductases. These oxidoreductases use NADPH for furfural reduction, depleting the available pools for biosynthesis. Thus furfural-mediated growth inhibition can be attributed to NADPH depletion [148], an insight that can be applied to other biocatalyst design projects.

PERSPECTIVES

Although many biocatalysts have been successfully redesigned for production of industrially important fuels and chemicals through traditional metabolic engineering, we are just beginning to see the potential of synthetic biology in this area. One of the foremost goals in our lab is the improvement of biocatalysts for biomass utilization. To attain this goal, tolerance to hydrolysate-derived inhibitors needs to be improved. For all applications, tolerance to high substrate and product titers is also important. This goal of redesigning a biocatalyst's phenotype, that is, tolerance, is not as clear as redesigning metabolism and a rational redesign strategy is particularly difficult when the mechanism of inhibition is not known.

As the understanding of our biocatalysts improves, particularly through reverse engineering of evolved strains, genome-scale models can be improved. Inclusion of kinetic and regulatory effects will also improve the accuracy and predictive power of these models. Note that some models have recently been developed that bypass the need for kinetic data, though [152]. Since enzymes are the major functional part performing the metabolic synthesis, improved protein engineering tools and new protein catalytic capability will aid in advancement of this field. It is important to generate high-quality protein mutagenesis libraries (relatively small libraries with a high diversity of enzymes) to facilitate efficient screening efforts [138]. Direct screening from metagenomic libraries of environmental samples can aid in isolation of enzymes with new functions, which cannot be obtained by the traditional strain isolation methods [153]. Enzymes can even be synthesized from scratch by a rational design strategy with computational aid [154]. Finally, new tools for better de novodesign of synthetic pathways need to be developed. Several databases, such as BNICE (Biochemical Network Integrated Computational Explorer) [155] and ReBiT (Retro-Biosynthesis Tool) [142], have already been established to facilitate identification of enzymes to construct a complete synthetic pathway for producing target compounds. It is important to establish guidelines, such as redox balance, energy production, and thermodynamic feasibility, to screen among these enormous pathways for the optimal routes.

By including synthetic biology tools in metabolic engineering projects, and vice versa, these two fields can significantly advance the replacement of petroleum-derived commodity products with those produced from biorenewable carbon and energy.

REFERENCES

1. L. R. Jarboe, T. B. Grabar, L. P. Yomano, K. T. Shanmugan, and L. O. Ingram, "Development of ethanologenic bacteria," Advances in Biochemical Engineering/Biotechnology, vol. 108, pp. 237–261, 2007.

2. D. G. Gibson, G. A. Benders, C. Andrews-Pfannkoch, et al., "Complete chemical synthesis, assembly, and cloning of a Mycoplasma genitalium genome," Science, vol. 319, no. 5867, pp. 1215–1220, 2008.

3. C. Laitigue, S. Vashee, M. A. Algire, et al., "Creating bacterial strains from genomes that have been cloned and engineered in yeast," Science, vol. 325, no. 5948, pp. 1693–1696, 2009.

4. J. E. Bailey, "Toward a science of metabolic engineering,"

Science, vol. 252, no. 5013, pp. 1668–1675, 1991.

5.   G. Stephanopoulos and J. J. Vallino, "Network rigidity and metabolic engineering in metabolite overproduction," Science, vol. 252, no. 5013, pp. 1675–1681, 1991.

6.   G. N. Stephanopoulos, A. A. Aristidou, and J. Nielsen, Metabolic Engineering: Principles and Methodologies, Academic Press, San Diego, Calif, USA, 1998.

7.   P. Ball, "Synthetic biology: starting from scratch," Nature, vol. 431, no. 7009, pp. 624–626, 2004.

8.   S. A. Benner and A. M. Sismour, "Synthetic biology," Nature Reviews Genetics, vol. 6, no. 7, pp. 533–543, 2005.

9.   R. McDaniel and R. Weiss, "Advances in synthetic biology: on the path from prototypes to applications,"Current Opinion in Biotechnology, vol. 16, no. 4, pp. 476–483, 2005.

10.   J. Pleiss, "The promise of synthetic biology," Applied Microbiology and Biotechnology, vol. 73, no. 4, pp. 735–739, 2006.

11.   L. Serrano, "Synthetic biology: promises and challenges," Molecular Systems Biology, vol. 3, article 158, 2007.

12.   J. D. Keasling, "Synthetic biology for synthetic chemistry," ACS Chemical Biology, vol. 3, no. 1, pp. 64–76, 2008.

13.   A. C. Forster and G. M. Church, "Synthetic biology projects in vitro," Genome Research, vol. 17, no. 1, pp. 1–6, 2007.

14.   D. Greber and M. Fussenegger, "Mammalian synthetic biology: engineering of sophisticated gene networks," Journal of Biotechnology, vol. 130, no. 4, pp. 329–345, 2007.

15.   W. Weber and M. Fussenegger, "The impact of synthetic biology on drug discovery," Drug Discovery Today, vol. 14, no. 19-20, pp. 956–963, 2009.

16.   C. E. French, "Synthetic biology and biomass conversion: a match made in heaven?" Journal of the Royal Society Interface, vol. 6, supplement 4, pp. S547–S558, 2009.

17.   S. K. Lee, H. Chou, T. S. Ham, T. S. Lee, and J. D. Keasling, "Metabolic engineering of microorganisms for biofuels

production: from bugs to synthetic biology to fuels," Current Opinion in Biotechnology, vol. 19, no. 6, pp. 556–563, 2008.

18. S. Picataggio, "Potential impact of synthetic biology on the development of microbial systems for the production of renewable fuels and chemicals," Current Opinion in Biotechnology, vol. 20, no. 3, pp. 325–329, 2009.

19. A. M. Feist, C. S. Henry, J. L. Reed, et al., "A genome-scale metabolic reconstruction for Escherichia coliK-12 MG1655 that accounts for 1260 ORFs and thermodynamic information," Molecular Systems Biology, vol. 3, article 121, 2007.

20. I. M. Keseler, C. Bonavides-Martínez, J. Collado-Vides, et al., "EcoCyc: a comprehensive view ofEscherichia coli biology," Nucleic Acids Research, vol. 37, supplement 1, pp. D464–D470, 2009.

21. D. Endy, "Foundations for engineering biology," Nature, vol. 438, no. 7067, pp. 449–453, 2005.

22. T. B. Grabar, S. Zhou, K. T. Shanmugam, L. P. Yomano, and L. O. Ingram, "Methylglyoxal bypass identified as source of chiral contamination in L(+) and D(-)-lactate fermentations by recombinantEscherichia coli," Biotechnology Letters, vol. 28, no. 19, pp. 1527–1535, 2006.

23. T. B. Causey, S. Zhou, K. T. Shanmugam, and L. O. Ingram, "Engineering the metabolism of Escherichia coli W3110 for the conversion of sugar to redox-neutral and oxidized products: homoacetate production," Proceedings of the National Academy of Sciences of the United States of America, vol. 100, no. 3, pp. 825–832, 2003.

24. K. Jantama, X. Zhang, J. C. Moore, K. T. Shanmugam, S. A. Svoronos, and L. O. Ingram, "Eliminating side products and increasing succinate yields in engineered strains of Escherichia coli C," Biotechnology and Bioengineering, vol. 101, no. 5, pp. 881–893, 2008.

25. L. P. Yomano, S. W. York, K. T. Shanmugam, and L. O. Ingram, "Deletion of methylglyoxal synthase gene (mgsA) increased

sugar co-metabolism in ethanol-producing Escherichia coli," Biotechnology Letters, vol. 31, no. 9, pp. 1389–1398, 2009.

26. P. C. Cirino, J. W. Chin, and L. O. Ingram, "Engineering Escherichia coli for xylitol production from glucose-xylose mixtures," Biotechnology and Bioengineering, vol. 95, no. 6, pp. 1167–1176, 2006.

27. X. Zhang, K. Jantama, J. C. Moore, K. T. Shanmugam, and L. O. Ingram, "Production of L-alanine by metabolically engineered Escherichia coli," Applied Microbiology and Biotechnology, vol. 77, no. 2, pp. 355–366, 2007.

28. R. S. Senger and E. T. Papoutsakis, "Genome-scale model for Clostridium acetobutylicum—part I: metabolic network resolution and analysis," Biotechnology and Bioengineering, vol. 101, no. 5, pp. 1036–1052, 2008.

29. K. R. Kjeldsen and J. Nielsen, "In silico genome-scale reconstruction and validation of the corynebacterium glutamicum metabolic network," Biotechnology and Bioengineering, vol. 102, no. 2, pp. 583–597, 2009.

30. J. Forster, I. Famili, P. Fu, et al., "Genome-scale reconstruction of the Saccharomyces cerevisiae metabolic network," Genome Research, vol. 13, no. 2, pp. 244–253, 2003.

31. M. R. Andersen, M. L. Nielsen, and J. Nielsen, "Metabolic model integration of the bibliome, genome, metabolome and reactome of Aspergillus niger," Molecular Systems Biology, vol. 4, article 178, 2008.

32. F. R. Blattner, et al., "The complete genome sequence of Escherichia coli K-12," Science, vol. 277, no. 5331, pp. 1453–1474, 1997.

33. R. U. Ibarra, J. S. Edwards, and B. O. Palsson, "Escherichia coli K-12 undergoes adaptive evolution to achieve in silico predicted optimal growth," Nature, vol. 420, no. 6912, pp. 186–189, 2002.

34. A. P. Bauer, S. M. Dieckmann, W. Ludwig, and K.-H. Schleifer, "Rapid identification of Escherichia colisafety and laboratory strain lineages based on Multiplex-PCR," FEMS Microbiology

Letters, vol. 269, no. 1, pp. 36–40, 2007.

35. S. Zhou, L. P. Yomano, K. T. Shanmugam, and L. O. Ingram, "Fermentation of 10% (w/v) sugar to D:(−)-lactate by engineered Escherichia coli B," Biotechnology Letters, vol. 27, no. 23-24, pp. 1891–1896, 2005.

36. M. Moniruzzaman, X. Lai, S. W. York, and L. O. Ingram, "Isolation and molecular characterization of high-performance cellobiose- fermenting spontaneous mutants of ethanologenic Escherichia coli KO11 containing the Klebsiella oxytoca casAB operon," Applied and Environmental Microbiology, vol. 63, no. 12, pp. 4633–4637, 1997.

37. G. Posfai, G. Plunkett III, T. Fehér, et al., "Emergent properties of reduced-genome Escherichia coli,"Science, vol. 312, no. 5776, pp. 1044–1046, 2006.

38. M. Durot, P.-Y. Bourguignon, and V. Schachter, "Genome-scale models of bacterial metabolism: reconstruction and applications," FEMS Microbiology Reviews, vol. 33, no. 1, pp. 164–190, 2009.

39. N. D. Price, J. A. Papin, C. H. Schilling, and B. O. Palsson, "Genome-scale microbial in silico models: the constraints-based approach," Trends in Biotechnology, vol. 21, no. 4, pp. 162–169, 2003.

40. P. Pharkya, A. P. Burgard, and C. D. Maranas, "OptStrain: a computational framework for redesign of microbial production systems," Genome Research, vol. 14, no. 11, pp. 2367–2376, 2004.

41. A. P. Burgard, P. Pharkya, and C. D. Maranas, "OptKnock: a bilevel programming framework for identifying gene knockout strategies for microbial strain optimization," Biotechnology and Bioengineering, vol. 84, no. 6, pp. 647–657, 2003.

42. H. Alper, Y.-S. Jin, J. F. Moxley, and G. Stephanopoulos, "Identifying gene targets for the metabolic engineering of lycopene biosynthesis in Escherichia coli," Metabolic Engineering, vol. 7, no. 3, pp. 155–164, 2005.

43. C. Bro and J. Nielsen, "Impact of 'ome' analyses on inverse

metabolic engineering," Metabolic Engineering, vol. 6, no. 3, pp. 204–211, 2004.

44. S. Y. Lee, D. Y. Lee, and T. Y. Kim, "Systems biotechnology for strain improvement," Trends in Biotechnology, vol. 23, no. 7, pp. 349–358, 2005.

45. T. Hermann, "Using functional genomics to improve productivity in the manufacture of industrial biochemicals," Current Opinion in Biotechnology, vol. 15, no. 5, pp. 444–448, 2004.

46. J. H. Choi, S. J. Lee, and S. Y. Lee, "Enhanced production of insulin-like growth factor I fusion protein inEscherichia coli by coexpression of the down-regulated genes identified by transcriptome profiling,"Applied and Environmental Microbiology, vol. 69, no. 8, pp. 4737–4742, 2003.

47. M.-J. Han, K. J. Jeong, J.-S. Yoo, and S. Y. Lee, "Engineering Escherichia coli for increased productivity of serine-rich proteins based on proteome profiling," Applied and Environmental Microbiology, vol. 69, no. 10, pp. 5772–5781, 2003.

48. B. F. Pfleger, D. J. Pitera, C. D. Smolke, and J. D. Keasling, "Combinatorial engineering of intergenic regions in operons tunes expression of multiple genes," Nature Biotechnology, vol. 24, no. 8, pp. 1027–1032, 2006.

49. S. A. Underwood, M. L. Buszko, K. T. Shanmugam, and L. O. Ingram, "Flux through citrate synthase limits the growth of ethanologenic Escherichia coli KO11 during xylose fermentation," Applied and Environmental Microbiology, vol. 68, no. 3, pp. 1071–1081, 2002.

50. M. Askenazi, E. M. Driggers, D. A. Holtzman, et al., "Integrating transcriptional and metabolite profiles to direct the engineering of lovastatin-producing fungal strains," Nature Biotechnology, vol. 21, no. 2, pp. 150–156, 2003.

51. J. Ohnishi, S. Mitsuhashi, M. Hayashi, et al., "A novel methodology employing Corynebacterium glutamicum genome information to generate a new L-lysine-producing

mutant," Applied Microbiology and Biotechnology, vol. 58, no. 2, pp. 217–223, 2002.

52. F. Martinez-Morales, A. C. Borges, A. Martinez, K. T. Shanmugam, and L. O. Ingram, "Chromosomal integration of heterologous DNA in Escherichia coli with precise removal of markers and replicons used during construction," Journal of Bacteriology, vol. 181, no. 22, pp. 7143–7148, 1999.

53. K. A. Datsenko and B. L. Wanner, "One-step inactivation of chromosomal genes in Escherichia coli K-12 using PCR products," Proceedings of the National Academy of Sciences of the United States of America, vol. 97, no. 12, pp. 6640–6645, 2000.

54. G. Pósfai, M. D. Koob, H. A. Kirkpatrick, and F. R. Blattner, "Versatile insertion plasmids for targeted genome manipulations in bacteria: isolation, deletion, and rescue of the pathogenicity island LEE of theEscherichia coli O157:H7 genome," Journal of Bacteriology, vol. 179, no. 13, pp. 4426–4428, 1997.

55. P. Gay, D. Le Coq, M. Steinmetz, et al., "Positive selection procedure for entrapment of insertion sequence elements in gram-negative bacteria," Journal of Bacteriology, vol. 164, no. 2, pp. 918–921, 1985.   ·

56. G. de la Cueva-Mendez and B. Pimentel, "Gene and cell survival: lessons from prokaryotic plasmid R1,"The EMBO Reports, vol. 8, no. 5, pp. 458–464, 2007.

57. S. Zhou, F. C. Davis, and L. O. Ingram, "Gene integration and expression and extracellular secretion of Erwinia chrysanthemi endoglucanase CelY (celY) and CelZ (celZ) in ethanologenic Klebsiella oxytoca P2," Applied and Environmental Microbiology, vol. 67, no. 1, pp. 6–14, 2001.

58. P. R. Jensen and K. Hammer, "The sequence of spacers between the consensus sequences modulates the strength of prokaryotic promoters," Applied and Environmental Microbiology, vol. 64, no. 1, pp. 82–87, 1998.

59. H. Alper, C. Fischer, E. Nevoigt, and G. Stephanopoulos,

"Tuning genetic control through promoter engineering," Proceedings of the National Academy of Sciences of the United States of America, vol. 102, no. 36, pp. 12678–12683, 2005.

60. I. Meynial-Salles, M. A. Cervin, and P. Soucaille, "New tool for metabolic pathway engineering inEscherichia coli: one-step method to modulate expression of chromosomal genes," Applied and Environmental Microbiology, vol. 71, no. 4, pp. 2140–2144, 2005.

61. C. Solem and P. R. Jensen, "Modulation of gene expression made easy," Applied and Environmental Microbiology, vol. 68, no. 5, pp. 2397–2403, 2002.

62. T. A. Carrier and J. D. Keasling, "Library of synthetic 5ʹ secondary structures to manipulate mRNA stability in Escherichia coli," Biotechnology Progress, vol. 15, no. 1, pp. 58–64, 1999.

63. Y. S. Park, S. W. Seo, S. Hwang, et al., "Design of 5ʹ-untranslated region variants for tunable expression in Escherichia coli," Biochemical and Biophysical Research Communications, vol. 356, no. 1, pp. 136–141, 2007.

64. F. J. Isaacs, D. J. Dwyer, C. Ding, D. D. Pervouchine, C. R. Cantor, and J. J. Collins, "Engineered riboregulators enable post-transcriptional control of gene expression," Nature Biotechnology, vol. 22, no. 7, pp. 841–847, 2004.

65. S. Jana and J. K. Deb, "Strategies for efficient production of heterologous proteins in Escherichia coli,"Applied Microbiology and Biotechnology, vol. 67, no. 3, pp. 289–298, 2005.

66. H. Alper, J. Moxley, E. Nevoigt, G. R. Fink, and G. Stephanopoulos, "Engineering yeast transcription machinery for improved ethanol tolerance and production," Science, vol. 314, no. 5805, pp. 1565–1568, 2006.

67. H. Alper and G. Stephanopoulos, "Global transcription machinery engineering: a new approach for improving cellular phenotype," Metabolic Engineering, vol. 9, no. 3, pp.

258–267, 2007.

68. I. P. Petrounia and F. H. Arnold, "Designed evolution of enzymatic properties," Current Opinion in Biotechnology, vol. 11, no. 4, pp. 325–330, 2000.

69. H. Zhao, L. Giver, Z. Shao, J. A. Affholter, and F. H. Arnold, "Molecular evolution by staggered extension process (StEP) in vitro recombination," Nature Biotechnology, vol. 16, no. 3, pp. 258–261, 1998.

70. A. Crameri, S.-A. Raillard, E. Bermudez, and W. P. C. Stemmer, "DNA shuffling of a family of genes from diverse species accelerates directed evolution," Nature, vol. 391, no. 6664, pp. 288–291, 1998.

71. C. Wang, M.-K. Oh, and J. C. Liao, "Directed evolution of metabolically engineered Escherichia coli for carotenoid production," Biotechnology Progress, vol. 16, no. 6, pp. 922–926, 2000.

72. P. C. Cirino and F. H. Arnold, "Protein engineering of oxygenases for biocatalysis," Current Opinion in Chemical Biology, vol. 6, no. 2, pp. 130–135, 2002.

73. H. H. Liao, "Alanine 2,3-aminomutase," US patent 7309597, 2007.

74. M. M. Altamirano, J. M. Blackburn, C. Aguayo, and A. R. Fersht, "Directed evolution of new catalytic activity using the / -barrel scaffold," Nature, vol. 403, no. 6770, pp. 617–622, 2000.

75. M. Leisola and O. Turunen, "Protein engineering: opportunities and challenges," Applied Microbiology and Biotechnology, vol. 75, no. 6, pp. 1225–1232, 2007.

76. A. Korkegian, M. E. Black, D. Baker, and B. L. Stoddard, "Computational thermostabilization of an enzyme," Science, vol. 308, no. 5723, pp. 857–860, 2005.

77. R. Machielsen, L. L. Looger, J. Raedts, et al., "Cofactor engineering of Lactobacillus brevis alcohol dehydrogenase by computational design," Engineering in Life Sciences, vol. 9, no. 1, pp. 38–44, 2009.

78. J. D. Bloom, M. M. Meyer, P. Meinhold, C. R. Otey, D. MacMillan, and F. H. Arnold, "Evolving strategies for enzyme engineering," Current Opinion in Structural Biology, vol. 15, no. 4, pp. 447–452, 2005.

79. R. A. Chica, N. Doucet, and J. N. Pelletier, "Semi-rational approaches to engineering enzyme activity: combining the benefits of directed evolution and rational design," Current Opinion in Biotechnology, vol. 16, no. 4, pp. 378–384, 2005.

80. O. Spadiut, I. Pisanelli, T. Maischberger, et al., "Engineering of pyranose 2-oxidase: improvement for biofuel cell and food applications through semi-rational protein design," Journal of Biotechnology, vol. 139, no. 3, pp. 250–257, 2009.

81. Q. Wang, A. R. Parrish, and L. Wang, "Expanding the genetic code for biological studies," Chemistry and Biology, vol. 16, no. 3, pp. 323–336, 2009.

82. Y. Lu, "Design and engineering of metalloproteins containing unnatural amino acids or non-native metal-containing cofactors," Current Opinion in Chemical Biology, vol. 9, no. 2, pp. 118–126, 2005.

83. J. C. Jackson, S. P. Duffy, K. R. Hess, and R. A. Mehl, "Improving nature›s enzyme active site with genetically encoded unnatural amino acids," Journal of the American Chemical Society, vol. 128, no. 34, pp. 11124–11127, 2006.

84. D. B. Walker, G. Joshi, and A. P. Davis, "Progress in biomimetic carbohydrate recognition," Cellular and Molecular Life Sciences, vol. 66, no. 19, pp. 3177–3191, 2009.

85. M. P. Cashion and T. E. Long, "Biomimetic design and performance of polymerizable lipids," Accounts of Chemical Research, vol. 42, no. 8, pp. 1016–1025, 2009.

86. S. S. Fong, A. P. Burgard, C. D. Herring, et al., "In silico design and adaptive evolution of Escherichia colifor production of lactic acid," Biotechnology and Bioengineering, vol. 91, no. 5, pp. 643–648, 2005.

87. X. Zhang, K. Jantama, J. C. Moore, L. R. Jarboe, K. T. Shanmugam, and L. O. Ingram, "Metabolic evolution of

energy-conserving pathways for succinate production in Escherichia coli," Proceedings of the National Academy of Sciences of the United States of America, vol. 106, no. 48, pp. 20180–20185, 2010.

88. J. G. Zeikus, M. K. Jain, and P. Elankovan, "Biotechnology of succinic acid production and markets for derived industrial products," Applied Microbiology and Biotechnology, vol. 51, no. 5, pp. 545–552, 1999.

89. J. B. McKinlay, C. Vieille, and J. G. Zeikus, "Prospects for a bio-based succinate industry," Applied Microbiology and Biotechnology, vol. 76, no. 4, pp. 727–740, 2007.

90. N. S. Samuelov, R. Lamed, S. Lowe, and I. G. Zeikus, "Influence of CO_2-HCO_3- levels and ph on growth, succinate production, and enzyme activities of Anaerobiospirillum succiniciproducens," Applied and Environmental Microbiology, vol. 57, no. 10, pp. 3013–3019, 1991.

91. M. J. Vander Werf, M. V. Guettler, M. K. Jain, and J. G. Zeikus, "Environmental and physiological factors affecting the succinate product ratio during carbohydrate fermentation by Actinobacillus sp. 130Z," Archives of Microbiology, vol. 167, no. 6, pp. 332–342, 1997.

92. P. Lee, S. Lee, S. Hong, and H. Chang, "Isolation and characterization of a new succinic acid-producing bacterium, Mannheimia succiniciproducens MBEL55E, from bovine rumen," Applied Microbiology and Biotechnology, vol. 58, no. 5, pp. 663–668, 2002.

93. D. G. Fraenkel, "Glycolysis," in Escherichia coli and Salmonella: Cellular and Molecular Biology, F. C. Neidhardt, Ed., ASM Press, Washington, DC, USA, 1996.

94. K. Jantama, M. J. Haupt, S. A. Svoronos, et al., "Combining metabolic engineering and metabolic evolution to develop nonrecombinant strains of Escherichia coli C that produce succinate and malate,"Biotechnology and Bioengineering, vol. 99, no. 5, pp. 1140–1153, 2008.

95. M.-K. Oh, L. Rohlin, K. C. Kao, and J. C. Liao, "Global

expression profiling of acetate-grown Escherichia coli," Journal of Biological Chemistry, vol. 277, no. 15, pp. 13175–13183, 2002.

96. K. C. Kao, L. M. Tran, and J. C. Liao, "A global regulatory role of gluconeogenic genes in Escherichia colirevealed by transcriptome network analysis," Journal of Biological Chemistry, vol. 280, no. 43, pp. 36079–36087, 2005.

97. R. Patnaik, W. D. Roof, R. F. Young, and J. C. Liao, "Stimulation of glucose catabolism in Escherichia coli by a potential futile cycle," Journal of Bacteriology, vol. 174, no. 23, pp. 7527–7532, 1992.

98. X. Zhang, K. Jantama, K. T. Shanmugam, and L. O. Ingram, "Reengineering Escherichia coli for succinate production in mineral salts medium," Applied and Environmental Microbiology, vol. 75, no. 24, pp. 7807–7813, 2009.

99. S. S. Ray and M. Bousmina, "Biodegradable polymers and their layered silicate nano composites: in greening the 21st century materials world," Progress in Materials Science, vol. 50, no. 8, pp. 962–1079, 2005.

100. A. K. Agrawal and R. Bhalla, "Advances in the production of poly(lactic acid) fibers. A review," Journal of Macromolecular Science-Polymer Reviews C, vol. 43, no. 4, pp. 479–503, 2003.

101. S. Benthin and J. Villadsen, "Production of optically pure D-lactate by lactobacillus bulgaricus and purification by crystallisation and liquid/liquid extraction," Applied Microbiology and Biotechnology, vol. 42, no. 6, pp. 826–829, 1995.

102. A. Demirci and A. L. Pometto, "Enhanced production of D(-)-lactic acid by mutants of lactobacillus-delbrueckii ATCC-9649," Journal of Industrial Microbiology, vol. 11, no. 1, pp. 23–28, 1992.

103. D. G. Fraenkel , "Glycolysis," in Escherichia coli and Salmonella: Cellular and Molecular Biology, F. C. Neidhardt, Ed., vol. 1, chapter 14, ASM Press, Washington DC, USA, 2nd

edition, 1996.

104. S. Zhou, T. B. Causey, A. Hasona, K. T. Shanmugam, and L. O. Ingram, "Production of optically pure D-lactic acid in mineral salts medium by metabolically engineered Escherichia coli W3110," Applied and Environmental Microbiology, vol. 69, no. 1, pp. 399–407, 2003.

105. S. Zhou, K. T. Shanmugam, L. P. Yomano, T. B. Grabar, and L. O. Ingram, "Fermentation of 12% (w/v)glucose to 1.2 M lactate by Escherichia coli strain SZ194 using mineral salts medium," Biotechnology Letters, vol. 28, no. 9, pp. 663–670, 2006.

106. M. Cheryan, S. Parekh, M. Shah, and K. Witjitra, "Production of acetic acid by Clostridium thermoaceticum," Advances in Applied Microbiology, vol. 43, pp. 1–33, 1997.   ·

107. C. Berraud, "Production of highly concentrated vinegar in fed-batch culture," Biotechnology Letters, vol. 22, no. 6, pp. 451–454, 2000.

108. S. N. Freer, "Acetic acid production by Dekkera/Brettanomyces yeasts," World Journal of Microbiology and Biotechnology, vol. 18, no. 3, pp. 271–275, 2002.

109. S. Y. Lee, J. H. Park, S. H. Jang, L. K. Nielsen, J. Kim, and K. S. Jung, "Fermentative butanol production by clostridia," Biotechnology and Bioengineering, vol. 101, no. 2, pp. 209–228, 2008.

110. L. D. Mermelstein, E. T. Papoutsakis, D. J. Petersen, and G. N. Bennett, "Metabolic engineering of Clostridium acetobutylicum ATCC 824 for increased solvent production by enhancement of acetone formation enzyme activities using a synthetic acetone operon," Biotechnology and Bioengineering, vol. 42, no. 9, pp. 1053–1060, 1993.

111. L. Harris, L. Blank, R. P. Desai, N. E. Welker, and E. T. Papoutsakis, "Fermentation characterization and flux analysis of recombinant strains of Clostridium acetobutylicum with an inactivated solR gene,"Journal of Industrial Microbiology and Biotechnology, vol. 27, no. 5, pp. 322–328, 2001.

112. R. V. Nair, E. M. Green, D. E. Watson, G. N. Bennett, and E. T. Papoutsakis, "Regulation of the sol locus genes for butanol and acetone formation in Clostridium acetobutylicum ATCC 824 by a putative transcriptional repressor," Journal of Bacteriology, vol. 181, no. 1, pp. 319–330, 1999.

113. R. P. Desai and E. T. Papoutsakis, "Antisense RNA strategies for metabolic engineering of Clostridium acetobutylicum," Applied and Environmental Microbiology, vol. 65, no. 3, pp. 936–945, 1999.

114. N. E. Altaras and D. C. Cameron, "Enhanced production of (R)-1,2-propanediol by metabolically engineered Escherichia coli," Biotechnology Progress, vol. 16, no. 6, pp. 940–946, 2000.

115. P. Soucaille, I. Meynial-Salles, F. Voelker, and R. Figge, "Microorganisms and methods for prodcution of 1,2-propanediol and acetol," WO, 2008, 2008/116853.

116. J. H. Park, K. H. Lee, T. Y. Kim, and S. Y. Lee, "Metabolic engineering of Escherichia coli for the production of L-valine based on transcriptome analysis and in silico gene knockout simulation,"Proceedings of the National Academy of Sciences of the United States of America, vol. 104, no. 19, pp. 7797–7802, 2007.

117. K. H. Lee, J. H. Park, T. Y. Kim, H. U. Kim, and S. Y. Lee, "Systems metabolic engineering of Escherichia coli for L-threonine production," Molecular Systems Biology, vol. 3, article 149, 2007.

118. L. O. Ingram, P. F. Gomez, X. Lai, et al., "Metabolic engineering of bacteria for ethanol production,"Biotechnology and Bioengineering, vol. 58, no. 2-3, pp. 204–214, 1998.

119. Y. Kim, L. O. Ingram, and K. T. Shanmugam, "Construction of an Escherichia coli K-12 mutant for homoethanologenic fermentation of glucose or xylose without foreign genes," Applied and Environmental Microbiology, vol. 73, no. 6, pp. 1766–1771, 2007.

120. L. O. Ingram, T. Conway, D. P. Clark, G. W. Sewell, and J.

F. Preston, "Genetic engineering of ethanol production in Escherichia coli," Applied and Environmental Microbiology, vol. 53, no. 10, pp. 2420–2425, 1987.

121. K. Ohta, D. S. Beall, J. P. Mejia, K. T. Shanmugam, and L. O. Ingram, "Genetic improvement ofEscherichia coli for ethanol production: chromosomal integration of Zymomonas mobilis genes encoding pyruvate decarboxylase and alcohol dehydrogenase II," Applied and Environmental Microbiology, vol. 57, no. 4, pp. 893–900, 1991.

122. L. P. Yomano, S. W. York, S. Zhou, K. T. Shanmugam, and L. O. Ingram, "Re-engineering Escherichia coli for ethanol production," Biotechnology Letters, vol. 30, no. 12, pp. 2097–2103, 2008.

123. A. Hasona, S. W. York, L. P. Yomano, L. O. Ingram, and K. T. Shanmugam, "Decreasing the level of ethyl acetate in ethanolic fermentation broths of Escherichia coli KO11 by expression of Pseudomonas putida estZ esterase," Applied and Environmental Microbiology, vol. 68, no. 6, pp. 2651–2659, 2002.

124. K. Hofvendahl and B. Hahn-Hagerdal, "Factors affecting the fermentative lactic acid production from renewable resources," Enzyme and Microbial Technology, vol. 26, no. 2–4, pp. 87–107, 2000.

125. S. D. Zhou, K. T. Shanmugam, and L. O. Ingram, "Functional replacement of the Escherichia coli D-(-)-lactate dehydrogenase gene (ldhA) with the L-(+)-lactate dehydrogenase gene (ldhL) from Pediococcus acidilactici," Applied and Environmental Microbiology, vol. 69, no. 4, pp. 2237–2244, 2003.

126. J. C. Parajó, H. Domínguez, and J. M. Domínguez, "Biotechnological production of xylitol—part 1: interest of xylitol and fundamentals of its biosynthesis," Bioresource Technology, vol. 65, no. 3, pp. 191–201, 1998.

127. T. Werpy and G. Petersen, Eds., Top Value Added Chemicals from Biomass: Volume I—Results of Screening for Potential Candidates from Sugars and Synthesis Gas, Pacific Northwest

National Laboratory and National Renewable Energy Laboratory, 2004.

128. T. B. Kim and D. K. Oh, "Xylitol production by Candida tropicalis in a chemically defined medium,"Biotechnology Letters, vol. 25, no. 24, pp. 2085–2088, 2003.

129. J. W. Chin, R. Khankal, C. A. Monroe, C. D. Maranas, and P. C. Cirino, "Analysis of NADPH supply during xylitol production by engineered Escherichia coli," Biotechnology and Bioengineering, vol. 102, no. 1, pp. 209–220, 2009.

130. P. Hols, M. Kleerebezem, A. N. Schanck, et al., "Conversion of Lactococcus lactis from homolactic to homoalanine fermentation through metabolic engineering," Nature Biotechnology, vol. 17, no. 6, pp. 588–592, 1999.

131. M. Ikeda, "Amino acid production processes," Advances in Biochemical Engineering/Biotechnology, vol. 79, pp. 1–35, 2003.

132. C. E. Nakamura and G. M. Whited, "Metabolic engineering for the microbial production of 1,3-propanediol," Current Opinion in Biotechnology, vol. 14, no. 5, pp. 454–459, 2003.

133. T. Hanai, S. Atsumi, and J. C. Liao, "Engineered synthetic pathway for isopropanol production inEscherichia coli," Applied and Environmental Microbiology, vol. 73, no. 24, pp. 7814–7818, 2007.

134. D.-K. Ro, E. M. Paradise, M. Quellet, et al., "Production of the antimalarial drug precursor artemisinic acid in engineered yeast," Nature, vol. 440, no. 7086, pp. 940–943, 2006.

135. Y. Yan, A. Kohli, and M. A. G. Koffas, "Biosynthesis of natural flavanones in Saccharomyces cerevisiae,"Applied and Environmental Microbiology, vol. 71, no. 9, pp. 5610–5613, 2005.

136. E. Leonard, Y. Yan, and M. A. G. Koffas, "Functional expression of a P450 flavonoid hydroxylase for the biosynthesis of plant-specific hydroxylated flavonols in Escherichia coli," Metabolic Engineering, vol. 8, no. 2, pp. 172–181, 2006.

137. E. Leonard, K. H. Lim, P.-N. Saw, and M. A. G. Koffas,

"Engineering central metabolic pathways for high-level flavonoid production in Escherichia coli," Applied and Environmental Microbiology, vol. 73, no. 12, pp. 3877–3886, 2007.

138. A. L. de Boer and C. Schmidt-Dannert, "Recent efforts in engineering microbial cells to produce new chemical compounds," Current Opinion in Chemical Biology, vol. 7, no. 2, pp. 273–278, 2003.

139. C. Schmidt-Dannert, D. Umeno, and F. H. Arnold, "Molecular breeding of carotenoid biosynthetic pathways," Nature Biotechnology, vol. 18, no. 7, pp. 750–753, 2000.

140. G. Sandmann, "Combinatorial biosynthesis of carotenoids in a heterologous host: a powerful approach for the biosynthesis of novel structures," ChemBioChem, vol. 3, no. 7, pp. 629–635, 2002.

141. M. Albrecht, S. Takaichi, S. Steiger, Z.-Y. Wang, and G. Sandmann, "Novel hydroxycarotenoids with improved antioxidative properties produced by gene combination in Escherichia coli," Nature Biotechnology, vol. 18, no. 8, pp. 843–846, 2000.

142. K. L. J. Prather and C. H. Martin, "De novo biosynthetic pathways: rational design of microbial chemical factories," Current Opinion in Biotechnology, vol. 19, no. 5, pp. 468–474, 2008.

143. S. Atsumi, T. Hanai, and J. C. Liao, "Non-fermentative pathways for synthesis of branched-chain higher alcohols as biofuels," Nature, vol. 451, no. 7174, pp. 86–89, 2008.

144. K. Zhang, M. R. Sawaya, D. S. Eisenberg, and J. C. Liao, "Expanding metabolism for biosynthesis of nonnatural alcohols," Proceedings of the National Academy of Sciences of the United States of America, vol. 105, no. 52, pp. 20653–20658, 2008.

145. R. Gonzalez, H. Tao, J. E. Purvis, S. W. York, K. T. Shanmugam, and L. O. Ingram, "Gene array-based identification of changes that contribute to ethanol tolerance in ethanologenic

Escherichia coli: comparison of KO11 (parent) to LY01 (resistant mutant)," Biotechnology Progress, vol. 19, no. 2, pp. 612–623, 2003.

146. L. P. Yomano, S. W. York, and L. O. Ingram, "Isolation and characterization of ethanol-tolerant mutants of Escherichia coli KO11 for fuel ethanol production," Journal of Industrial Microbiology and Biotechnology, vol. 20, no. 2, pp. 132–138, 1998.

147. M. P. Brynildsen and J. C. Liao, "An integrated network approach identifies the isobutanol response network of Escherichia coli," Molecular Systems Biology, vol. 5, article 277, 2009.

148. E. N. Miller, et al., "Silencing of NADPH-dependent oxidoreductase genes (yqhD and dkgA) in furfural-resistant ethanologenic Escherichia coli," Applied and Environmental Microbiology, vol. 75, no. 13, pp. 4315–4323, 2009.

149. A. Martinez, M. E. Rodriguez, M. L. Wells, S. W. York, J. F. Preston, and L. O. Ingram, "Detoxification of dilute acid hydrolysates of lignocellulose with lime," Biotechnology Progress, vol. 17, no. 2, pp. 287–293, 2001.

150. R. Gonzalez, H. Tao, K. T. Shanmugam, S.W. York, and L. O. Ingram, "Transcriptome analysis of ethanologenic Escherichia coli strains: tolerance to ethanol," in Proceedings of the 225th ACS National Meeting, p. U200, New Orleans, La, USA, March 2003.

151. E. N. Miller, L. R. Jarboe, P. C. Turner, et al., "Furfural inhibits growth by limiting sulfur assimilation in ethanologenic Escherichia coli strain LY180," Applied and Environmental Microbiology, vol. 75, no. 19, pp. 6132–6141, 2009.

152. L. M. Tran, M. L. Rizk, and J. C. Liao, "Ensemble modeling of metabolic networks," Biophysical Journal, vol. 95, no. 12, pp. 5606–5617, 2008.

153. P. D. Schloss and J. Handelsman, "Biotechnological prospects from metagenomics," Current Opinion in Biotechnology, vol. 14, no. 3, pp. 303–310, 2003. · ·

154. L. Jiang, E. A. Althoff, F. R. Clemente, et al., "De novo computational design of retro-aldol enzymes,"Science, vol. 319, no. 5868, pp. 1387–1391, 2008.

155. V. Hatzimanikatis, C. Li, J. A. Ionita, C. S. Henry, M. D. Jankowski, and L. Broadbelt, "Exploring the diversity of complex metabolic networks," Bioinformatics, vol. 21, no. 8, pp. 1603–1609, 2005.

10

Analysis of a Recent Biofilter Model for Toluene Biodegradation

Department of Chemical Engineering, American University of Sharjah, Sharjah, UAE

ABSTRACT

This paper investigates and provides a critical analysis of the toluene biofilter model developed by Li and De Visscher. The model simulation results have been reproduced and compared with several sets of experimental data from literature. Three different model variations are considered: model with no substrate inhibition, with substrate inhibition, and with air flow rate modification. A sensitivity analysis has been performed on model to study the effect of important parameters on the removal efficiency. Model limitations and improvements have been highlighted.

INTRODUCTION

The upsurge of strict environmental regulations to maintain good air, soil and water quality requires the use of proper pollution control and pollution prevention equipment. Traditional pollution control technologies such as incineration, carbon adsorption and wet-scrubbing may be used to control pollution. However, these conventional techniques are becoming more expensive due to more stringent environmental regulations [1]. In the field of pollution control, biological treatment such as biofiltration is continuing to gain attention as an alternative to the conventional techniques.

Biofiltration is a pollution control technique that uses living microorganisms to capture and degrade pollutants from air. It is a process that combines basic mechanisms of adsorption, biodegradation and desorption of gas phase pollutants [2]. A biofilter is simply a packed bed that utilizes a packing material with microorganisms such as bacteria immobilized as biofilm on the surface and the pore structures of the packing material. The packing material may be particles such as peat, compost, peat/perlite mixture or organic or inorganic commercial media materials. The flow of contaminated air through the biofilter results in the degradation of the pollutants by the immobilized microorganisms. Biofilter performance is affected by a number of factors such as the composition and relative humidity of the waste stream, airflow velocity, temperature and pH of the biofilter bed, the pore size distribution, and other structural characteristics [1]. The surface of the porous material in the biofilter is covered with biofilms which are made of microorganisms. Treatment begins with the transfer of the contaminants from the air stream to the biofilm phase. Then, the dissolved contaminant is moved by diffusion and by advection in the air. Biotransformation finally converts the contaminant to biomass, metabolic by-products, carbon dioxide and water.

Several types of biofilters have been developed, the most typical of which include bioscrubbers, trickling bed biofilter and packed bed biofilter [3]. In all these technologies, pollutants in the gas stream are transferred to the biofilm and are degraded by the microorganisms. The bioscubber consists of a scrubber and a bioreactor. In the scrubber, water is sprayed counter-current to the polluted gas flow resulting in absorption of the pollutant into the water. This water is then directed to a bioreactor containing activated sludge where microorganisms degrade the pollutants. The trickling bed biofilter, on the other hand, relies on the inert packing media to support bacterial growth. The packed bed biofilter does not use a large continuous flow of water. The media used in the packed bed biofilter acts as a water reservoir as well as a support structure for the bacteria.

Figure 1 shows a simplified schematic of a biofilter [3].

REVIEW ON BIOFILTER MODELS

The design and scale-up of biofilters requires development of realistic mathematical models [4]. Many mathematical biofilter models have been developed in an effort to improve our understanding of biofilters, to guide experimentation, and to improve the biofilter design, performance and scale-up [5].

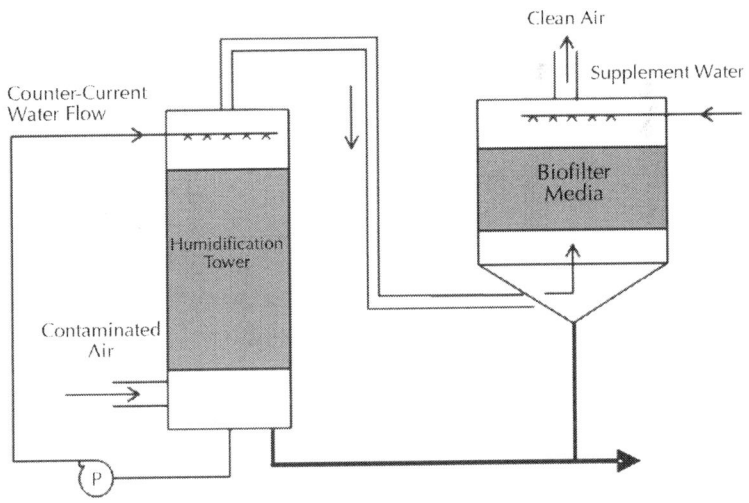

Figure 1: Simplified biofilter schematic.

Most of these models are based on several assumptions and may be simple or complex models. Over the years, researchers have developed more and more complex biofilter models that accurately describe the biofilter performance and provide a more rational approach to biofilter design. While there has been significant success among investigators in describing and understanding laboratory results of biofilters, no single model has become of a standard that is generally accepted [5]. Different models have been developed that use different approaches to describe the pollutant degradation, biofilm growth and biofilter performance. Some of the biofilter models are very simple and consider no growth or mass-transfer limitations [6].

Ottengraf and van der Oever [7] developed the first steady-state biofilter model for the biofiltration of single volatile organic compounds (VOC) in 1980s. In this early model, both diffusion and biodegradation of pollutants in the biofilm were considered. In particular, Ottengraf and van der Oever considered three cases in their model: first-order kinetics, zero-order kinetics with reaction rate limiting, and zero-order kinetics with diffusion rate limiting. The basic assumptions in the model included smaller biofilm thickness compared to the diameter of support particles and that the gas phase is in plug flow. Ottengraf and van der Oever model is the most widely used biofilter model since it is easy to implement.

Later Shareefdeen et al. [8] extended the work of Ottengraf and proposed a biofilter model to describe the steady-state performance of biofilters. This model includes both oxygen and substrate inhibition effects and oxygen is not assumed to be limiting as in the case of the model by Ottengraf and van der Oever [7]. Later, Shareefdeen and Baltzis [9] modified the model by Shareefdeen et al. [8] and proposed the partial coverage of the support particles by biofilms, leaving the bare surfaces of the particles in direct contact with the airstream. Furthermore, Shareefdeen and Baltzis proposed that the uncovered surfaces can adsorb the substrate, and that the adsorption follows the Freundlich isotherm. These two models have a wide range of applicability but require input data that are not easily available.

Similar to the model of Shareefdeen et al. [8], Deshusses et al. [10, 11] developed a dynamic diffusionreaction model to describe the steady-state as well as transient behavior of biofilters. This model considers a first order exchange rate between biomass layers and also includes the substrate inhibition effects.

Several biofilter models have been developed in recent years. Almost all of these models have been developed based on these earlier biofilter models. The biofilter model by Deshusses et al. was modified in 2006 by Park and Jung [12] by using Luong kinetics for substrate inhibition instead of Michaelis-Menten kinetics. This model is used to study biofiltration of toluene under high load conditions.

Yang et al. [13] developed a new biofilter model for the biofiltration of toluene in a rotating drum biofilter. This model takes into account a variable biofilm thickness. According to this model, the biofilm thickness increases by microbial growth. As a result, this model becomes rather complicated.

Babu and Raghuvanshi [2] proposed a transient mathematical model for biofilter operation and biofiltartion of VOCs in periodic mode. Under transient conditions, the uncovered surface of the solid support in the biofilter can adsorb the pollutants. Therefore, in this model, adsorption process is accounted explicitly. The model assumes axial dispersion flow for the gas and a linear driving force to approximate the pollutant or contaminant interphase transport. It also assumes pollutant adsorption and biodegradation in the solid support using first order kinetics with no oxygen limitation.

Liao et al. [14] developed a mathematical model for gas-liquid two phase flow and biodegradation in a trickling biofilter subjected to low concentration of VOC. The model simplifies the packed biofilter as a series of capillary tubes covered by the biofilm. The model assumes that the biofilm is formed on the exterior surface of the packed material and thus no reaction occurs in the pores of the packed material. The absorption of the VOC pollutant at the gas-liquid interface is evaluated using Henry's law. In short, the model incorporates the effect of pollutant adsorption at gas-liquid interface, the mass transfer resistance in the liquid zone and the biofilm zone, the biochemical reaction in the biofilm, and the limitation of oxygen for the microbial growth.

Chen et al. [15] described a model for removal of nitric oxide in a rotating drum biofilter. The model is based on mass balance of the pollutant in the gas, liquid, and biofilm phases. Based on the mass component profile of NO at the gas-liquid interface combined with Monod kinetic equation, the model predicts the mass transfer-reaction process of NO in the rotating bed biofilter.

The mathematical model developed by Spigno et al. [16] for phenol degradation in biofilters assumes no oxygen limitation, no reaction inside the pores of the solid material, fast gas-solid transfer compared to diffusion and reaction in the biofilm, uniform biofilm properties, and Monod kinetics to describe the biofilm growth.

Biofiltration of ammonia has been recently studied by Baquerizo et al. [17]. The mathematical model is based on discretized mass balances and nitrification kinetics that include inhibitory effects caused by free nitrous acid and free ammonia in the biofilter. The model includes most of the known biofiltration phenomenon and accounts for advection, absorption, adsorption, diffusion and biodegradation. In addition, all biological inhibitions occurring in the nitrification process and oxygen limitation in the kinetic model have been considered. In general, the model is able to predict ammonia shock-loadings and the biofilter behavior under inhibitory conditions.

Lu et al. [18] described the biofiltration of isopropyl alcohol and acetone mixtures assuming pseudo-steadystate operating conditions. The model neglects the convective transport within the biofilm. Chimel et al. [19] developed a concise model for VOC removal in periodic operation of biofilters. Under periodic operation, the biofilm is assumed to be less developed than under continuous regime, due to temporal nutrient shortages. A linear driving force is used to approximate the pollutant interphase transport. The model also allows for the pollutant adsorption and biodegradation in the support phase using the first-order kinetics with no oxygen limitation.

Toluene is one of the most widely released air pollutants. It is used extensively in fuels, solvents and as raw material for production of other chemicals [20]. Studies in humans and animals have demonstrated that toluene is readily absorbed via the lungs and the gastrointestinal tract. In humans, exposure to toluene causes central nervous system depression and may also act as a narcotic in case of large dosage. Chronic occupational exposure and incidences of toluene abuse have resulted in heaptomegaly and liver function changes [21]. Thus, effective removal of toluene from contaminated gas and liquid streams becomes important.

The objective of this paper is to analyze and investigate the validity of the model of Li and De Visscher [22] for biofiltration of toluene contaminated air. This model is based on the model by De Visscher and Van Cleemput [23] for methane biodegradation in landfill cover soils. Li and De Visscher [22] used the model by De Visccher and Van Cleepmput [23] to study toluene biofiltration, and modified the model equation to include substrate inhibittion using Haldane kinetics and influence of gas flow rate on toluene degrading activity.

TOLUENE BIOFILTER MODEL OF LI AND DE VISSCHER [22]

In this section the biofilter model of Li and De Visscher [22] is presented so that the readers can follow the analysis and discussion presented in the subsequent sections.

In the toluene biofilter model by Li and De Visscher, the biofilter is considered as a plug flow reactor. The height of the biofilter is divided into a number of subsections and the mass balance in each of these subsections is given as follows:

$$Q \cdot C_j = Q \cdot C_{j+1} + r_j A \Delta z \tag{1}$$

where Q is the air flow rate (m³/h), A is cross-sectional area of the biofilter (m²), C_j and C_{j+1} are concentrations (g/m³) of the gas-phase pollutant (toluene) in the subsections j and j + 1 respectively, Δz is the subsection height (m), and r_j is the volumetric biodegradation rate (g/m³biofilter/h) in the subsection j of the biofilter.

Equation (1) can be written in the following differential form with the biofilter height z as the independent variable:

$$\frac{dC}{dz} = -\frac{A}{Q}r$$

(2)

The expression for the volumetric biodegradation rate r depends on whether substrate inhibition is considered or not. On the basis of Pirt kinetics [24] and logistic growth rate expression, De Visscher and Van Cleemput [23] developed the following expression for microbial growth for methane biofiltration:

$$\mu = \mu'_{max} \frac{\left[1 - \frac{V_{max}}{V_{max,max}}\right]S}{K_m + S} - a$$

(3)

where V_{max} is the maximum degradation rate (g pollutant/m³ biofilter/h), $V_{max,max}$ is the maximum attainable value of V_{max} (g pollutant/m³ biofilter/h), S is the substrate concentration in the liquid phase (g/m³), K_m is the Michaelis-Menten constant (g/m³), a is the decay rate of the microbes (h⁻¹), μ is the specific growth rate (h⁻¹), and μ_{max} is the actual maximum specific growth rate (h⁻¹). The change of V_{max} with time is given by the following equation:

$$\frac{dV_{max}}{dt} = \mu V_{max}$$

(4)

De Visscher and Van Cleemput [23] proposed that V_{max} attains a maximum value $V_{max,max}$ during the course of biofiltration. Most biofilter models assume a uniform biofilm structure and a constant V_{max} throughout the biofiltration process. However, in reality, the thickness of bio-layer changes with substrate concentration and thus changes with the position in the biofilter [25]. The biofilm is expected to grow the thickest where the substrate concentration is the highest. Therefore, V_{max} is not constant along the biofilter height.

Biofilter Model 1 (without Substrate Inhibition)

If substrate inhibition is not considered, the volumetric biodegradation rate r_j is given by Michaelis-Menten kinetics:

$$r_j = V_{max,j} \frac{S_j}{K_m + S_j} \tag{5}$$

where S_j is the liquid-phase concentration (g/m^3) of the pollutant and is related to C_j by Henry's law as follows:

$$S_j = \frac{C_j}{H_{cc}} \tag{6}$$

where H_{cc} is the dimensionless Henry's constant for the pollutant. In each subsection, the microbial growth rate is given by De Visscher and Van Cleemput [23] model:

$$\mu_j = \mu'_{max} \frac{\left[1 - \frac{V_{max,j}}{V_{max,max}}\right] S_j}{K_m + S_j} - a \tag{7}$$

At steady state, there is no net microbial activity ($\mu_j = 0$) and Equation (7) can be written as follows:

$$V_{max,j} = \left[1 - \frac{a}{\mu'_{max}}\left(1 + \frac{K_m}{C_j / H_{cc}}\right)\right] V_{max,max} \tag{8}$$

Equation (8) indicates that the V_{max} is not constant but changes with the pollutant concentration at different biofilter height positions. Equation (2) combined with Equation (5) and Equation (8) can be solved numerically to study the pollutant concentration along the biofilter height.

Biofilter Model 2 (with Substrate Inhibition)

In case of substrate inhibition, Haldane kinetics [26] gives the following expression for the microbial growth rate:

$$\mu_j = \mu'_{max} \frac{\left[1 - \dfrac{V_{max,j}}{V_{max,max}}\right] S_j}{K_m + S_j + \dfrac{S_j^2}{K_I}} - a$$

(9)

where K_I is the inhibition constant (g/m^3). Similarly, the volumetric biodegradation rate assumes the following form:

$$r_j = V_{max,j} \frac{S_j}{K_m + S_j + \dfrac{S_j^2}{K_I}}$$

(10)

Again, under steady-state, μ_j becomes equal to zero and Equation (9) becomes as follows:

$$V_{max,j} = \left[1 - \frac{a}{\mu'_{max}}\left(1 + \frac{\dfrac{K_m}{C_j}}{H_{cc}} + \frac{C_j}{H_{cc}K_I}\right)\right] V_{max,max}$$

(11)

In case of substrate inhibition, Equation (2) combined with Equations (10) and (11) can be solved numerically to study the pollutant concentration along the biofilter height.

Biofilter Model 3 (with Flow Rate Modification)

The performance of a biofilter is significantly affected by the flow rate of air. At high air flow rates, the biofilm layer becomes thinner and more uniform [27], and mass transfer and biodegradation are favored [22]. For biofilters operating under varying air flow rates with substrate inhibition (model 2), an empirical equation is used to describe the effect of air flow rate Q on the value of $V_{max,max}$ [22]:

$$V'_{max,max} = V_{max,max} \, Q^p \tag{12}$$

Where $V_{max,max}$ is the flow-rate-modified maximum value of $V_{max,max}$. The value of p is between 0 and 1 [22]. To include the effect of varying air flow rate, Equation (12) replaces $V_{max,max}$ in Equation (11) of model 2. The equations can then be solved to study the concentration of pollutant along the biofilter height.

RESULTS AND DISCUSSION

Biofilter Model 1 (without Substrate Inhibition)

The non-linear coupled differential equations made up of Equations (2), (5) and (8), are solved using the ODE command in MATLAB to describe the variation of gasphase toluene concentration along the biofilter height for the case where substrate inhibition is neglected.

Model 1: Verification Using the Data of Zamir et al. [28]

Li and De Visscher [22] validated their model with the experimental results of Aizpuru et al. [29]. These results are also reproduced and are in good agreement as reported by Li and De Visscher [22]. The model of Li and De Visscher [22] was also solved and verified using the data of Zamir et al. [28]. Fungi were used as the toluenedegrading microorganisms. The biofilter had an inner diameter of 9.9 cm, a total height of 75 cm and an effecttive volume of 4 L (empty basis) and consisted of three stages. A mixture of powdered compost supplied acted as packing material. The parameters required for model 1 solution are summarized in Table 1.

Since fungi were used for biodegradation, substrate inhibition can be neglected and model 1 can be used to predict the gas-phase toluene concentration along the biofilter height. Figure 2 shows a good agreement between the model simulation results and the experimental results adopted from Zamir et al. [28].

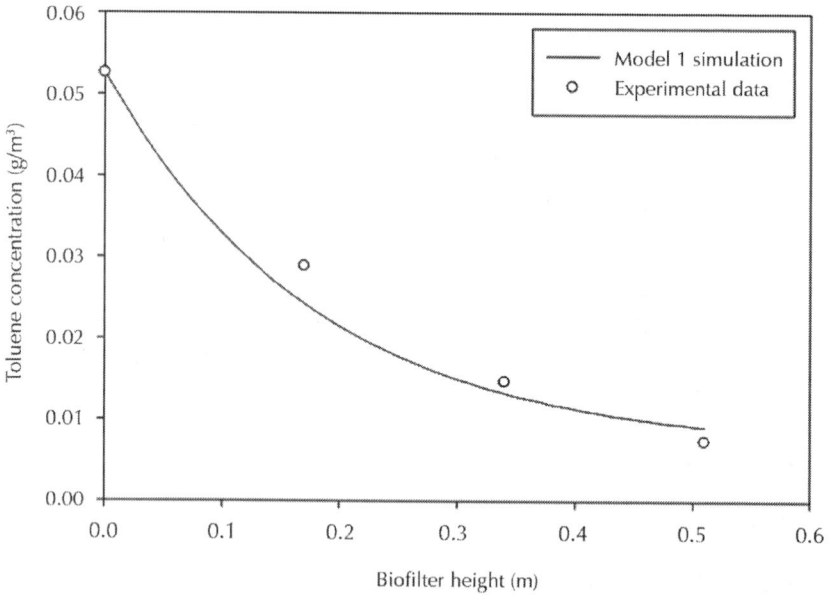

Figure 2: Model 1 (no substrate inhibition) prediction of toluene gas-phase concentrations along the biofilter height. Toluene-contaminated air flow rate is 0.06 m³/h, inlet toluene gas-phase concentration is 0.0526 g/m³, experimental data adopted from Zamir et al. [28].

Table 1: Model 1 parameters (no substrate inhibition) [28]

Parameter	Value	Unit	Source
μ_{max}¢	0.28	h⁻¹	[30,31]
K_m	3.495	g/m³	[22]

a	0.0017	h^{-1}	[32]
$V_{max,max}$	50	g pollutant/m³ biofilter/h	[32]
H_{cc}	0.313	-	[33]

Therefore, in cases where substrate inhibition can be neglected, such as use of fungi for biodegradation or low toluene loading, model 1 developed by Li and De Visscher [22] showed excellent agreement with the two sets different experimental data as described by Li and De Visscher [22] and also Figure 2.

Biofilter Model 2 (with Substrate Inhibition)

When the substrate loading is very high or the microorganisms used for biodegradation are not fungi, substrate inhibition needs to be taken into account. This corresponds to model 2 developed earlier. In this case, the Equation (2) combined with Equations (10) and (11) can be solved to study the variation of gas-phase toluene concentration along the biofilter height.

Model 2: Verification using the Data of Park and Jung [12]

The model 2 is solved, reproduced and compared with experimental results of Park and Jung [12].Table 2 shows the parameters which were used to solve the biofilter model as per Park and Jung [12].

The experiments conducted by Park and Jung [12] for toluene biofiltration were carried out in three biofilters in series, each packed with spherical ceramic packing. The total packing height was 0.54 m. Pseudomonas pudita F1 stains were used as toluene-degrading microorganisms. A volumetric loading rate of 35.6 m³ air/m³ biofilter/h was used. Figure 3 shows the solution of model 2 with substrate inhibition and comparison with the experimental results. Reproduced model solutions and experimental data are in good agreement.

To verify the importance of inhibition effects, the model 1 (substrate with no inhibition model) is solved and compared with the same set of experimental data of Park and Jung [12]. Figure 4shows the simulation results using model 1 and parameters in Table 2 in case if the substrate inhibition were neglected. Figure 4 clearly indicates that neglecting the substrate inhibition gives poor agreement.

In the case of Park and Jung [12], Pseudomonas pudita F1 stains were used as toluene-degrading microorganisms and substrate inhibition becomes important.

Table 2: Model 2 parameters (including substrate inhibition) [12]

Parameter	Value	Unit	Source
μ_{max} ¢	0.58	h^{-1}	[34-38]
K_m	5.34	g/m$_3$	[22]
a	0.0017	h^{-1}	[32]
$V_{max,max}$	123.1	g pollutant/m^3 biofilter/h	[22]
K_I	2.66	g/m^3	[22]
H_{cc}	0.276	-	[33]

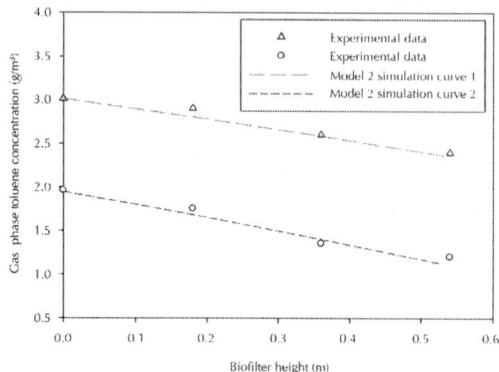

Figure 3: Model 2 (with substrate inhibition) prediction of toluene gas-phase concentrations along the biofilter height. Toluene-contaminated

air volumetric loading is 35.6 m³/m³ biofilter/h, curve 1: Inlet toluene gas-phase concentration is 3.01 g/m³, curve 2: Inlet toluene gas-phase concentration is 1.96 g/m³, experimental data adopted from Park and Jung [12].

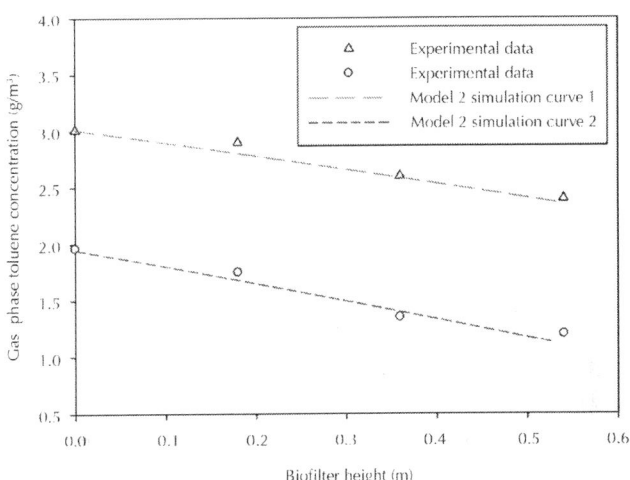

Figure 4: Model 1 (no substrate inhibition) prediction of toluene gas-phase concentrations along the biofilter height. Toluene-contaminated air volumetric loading is 35.6 m³/m³ biofilter/h, curve 1: Inlet toluene gas-phase concentration is 3.01 g/m³, curve 2: Inlet toluene gas-phase concentration is 1.96 g/m³, experimental data adopted from Park and Jung [12].

On the other hand, the experiments by Aizpuru et al. [29] used fungi as the toluene-biodegrading microorganisms for which substrate inhibition was negligible [22]. These analyses demonstrate careful selection of the model and the kinetics of biodegradation.

Biofilter Model 3 (with Flow Rate Modification)

Model 3 developed by Li and De Visscher [22] accounts for the effect of varying air flow rate. In this case, Equation (2) combined with Equations (10)-(12) must be solved simultaneously. The model solutions are reproduced and were compared with experimental results of Vergara-Fernández et al. [38]. The parameters used to solve model 3 are summarized in Table 3.

Their experiments were carried out at 32°C - 37°C in a toluene-degrading biofilter subjected to varying air flow rates. Compost and seashells were used as a packing material with microbial flora present in the compost. The height and diameter of the biofilter were 0.75 m and 0.145 m, respectively. Inhibition is considered in this case [39]. The solution to model 3 using the parameters in Table 3 are obtained for four different flow rates 0.12, 0.18, 0.24 and 0.73 m³/h.Figure 5 shows the simulation results with the experimental data from VergaraFernández et al. [38] for the flow rate of 0.12 m³/h. On the same graph, experimental results were also compared with model 2 which neglect the effect of air flow rate variation. Model 2 shows wider variation with the experimental data than the simulation results presented by Li and De Visscher [22]. This indeed confirms that flow rate modification is an important factor which should not be ignored.

Sensitivity Analysis

Effect of Actual Maximum Growth Rate (μ'_{max})

A sensitivity analysis of model 1 - 3 is performed to study the effect of the actual maximum specific growth rate (μ'_{max}) on the removal efficiency of toluene. The effect of changing (μ'_{max}), with all other parameters fixed as in Table 1 is shown in Figure 6. An increase in (μ'_{max}) leads to an increase in the removal efficiency of toluene in case of model 1. However, percent removal efficiency (%RE) asymptotically reaches a maximum value at (μ'_{max}) of 0.70 h⁻¹. In case of model 2 (with substrate inhibition), the effect of changing (μ'_{max}), while keeping other parameters in Table 2 constant, is shown in Figure 7. Again, an increase in the maximum specific growth rate increases the removal efficiency up to a certain limit, after which the change is negligible. Since model 3 is gas-phase concentration is 4.6 g/m³, $\mu'_{max,experimental}$ is 0.28 h⁻¹.

very similar to model 2, the effect of changing the actual maximum specific growth rate will be the same as in case of model 2.

Table 3: Model 3 parameters (with flow rate modification) [38]

Parameter	Value	Unit	Source
μ_{max}	0.12	h⁻¹	[39]
K_m	11.1	g/m³	[22]
a	0.0017	h⁻¹	[32]
K_l	95.5	g/m³	[22]
$V_{max,max}$	869.3	g pollutant/m³ biofilter/h	[22]
p	0.6325	-	[22]
H_{cc}	0.422	-	[33]

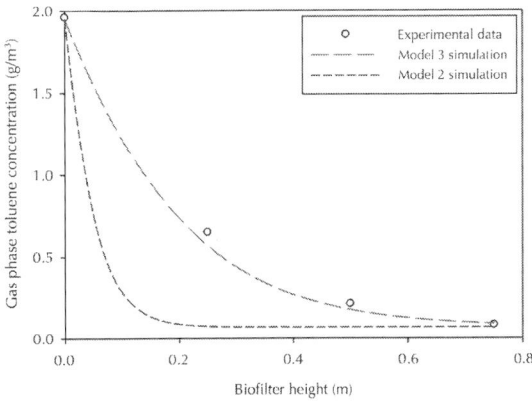

Figure 5: Model 3 (with flow rate modification) prediction of toluene gas-phase concentrations along the biofilter height. Toluene-contaminated air volumetric flow rate is 0.12 m³/h, inlet toluene gas-phase concentration is 1.95 g/m³, experimental data adopted from Vergara-Fernándaz et al. [38].

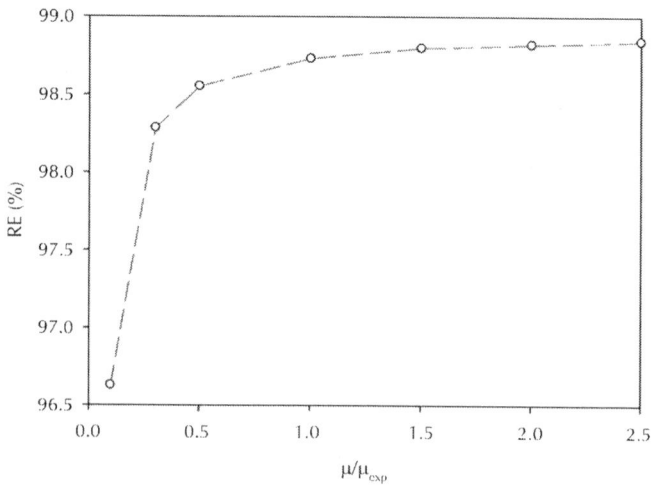

Figure 6: Effect of changing (μ'_{max}) in model 1 (no substrate inhibition) on removal efficiency (RE). Toluene-contaminated air volumetric loading is 77.6 m³/m³ biofilter/h, inlet toluene.

Effect of Michaelis-Menten Constant (K_m)

The effect of changing the Michaelis-Menten constant (K_m) on the removal efficiency was studied. In case of model 1, all parameters in Table 1 were fixed and the Michaelis-Menten constant was changed. The experimental K_m value was 1.24 g/m³. The effect of changing K_m on removal efficiency in model 1 is depicted in Figure 8. For model 1 that neglects substrate inhibition, an increase in the value of the Michaelis-Menten constant (K_m) results in a decrease in removal efficiency. The same trend was observed at different inlet toluene gasphase concentrations. In case of model 2 with substrate inhibition, the effect of changing K_m, while keeping other parameters in Table 2 constant, is shown in Figure 9 for inlet toluene gas-phase concentration is 3.01 g/m³. Again, an increase in the value of the Michaelis-Menten constant (K_m) results in a decrease in removal efficiency. Since model 3 is very similar to model 2, the effect of changing K_m will be the same as in case of model 2.

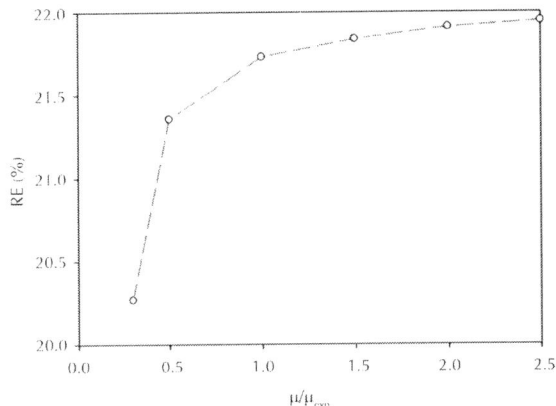

Figure 7: Effect of changing (μ'_{max}) in model 2 (with substrate inhibition) on removal efficiency (RE). Toluene-contaminated air volumetric loading is 35.6 m³/m³ biofilter/h, inlet toluene gas-phase concentration is 3.01 g/m³, $\mu'_{max,experimental}$ is 0.58 h⁻¹.

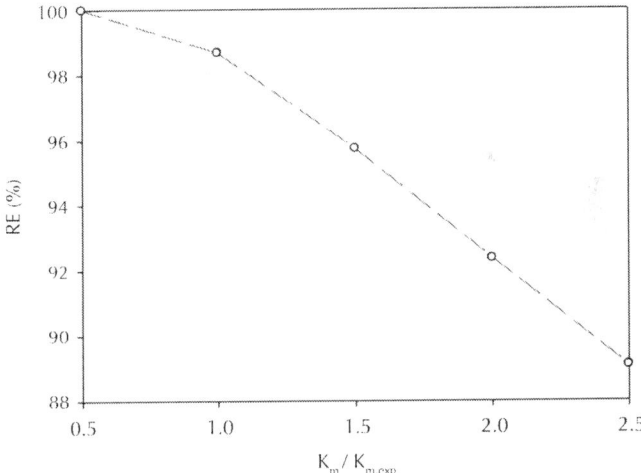

Figure 8: Effect of changing K_m in model 1 (no substrate inhibition) on removal efficiency (RE). Toluene-contaminated air volumetric loading is 77.6 m³/m³ biofilter/h, inlet toluene gas-phase concentration is 4.6 g/m³, $K_{m,experimental}$ is 1.24 g/m³.

Effect of Decay Rate (a)

The effect of changing the decay rate on removal efficiency in model 1 is shown in Figure 10. The other pa- rameters were kept constant as in Table 1. As expected, the higher the decay rate of the biomass, the lower the removal efficiency in case of model 1. Similarly, the effect of changing the decay rate on removal efficiency in model 2 is shown in Figure 11. Model 3 also gave similar results. In all cases, for the range of decay rates selected, the effect on the removal efficiency was not very significant. Nevertheless, the conditions of the biofilter must be properly selected in order to minimize biomass decay.

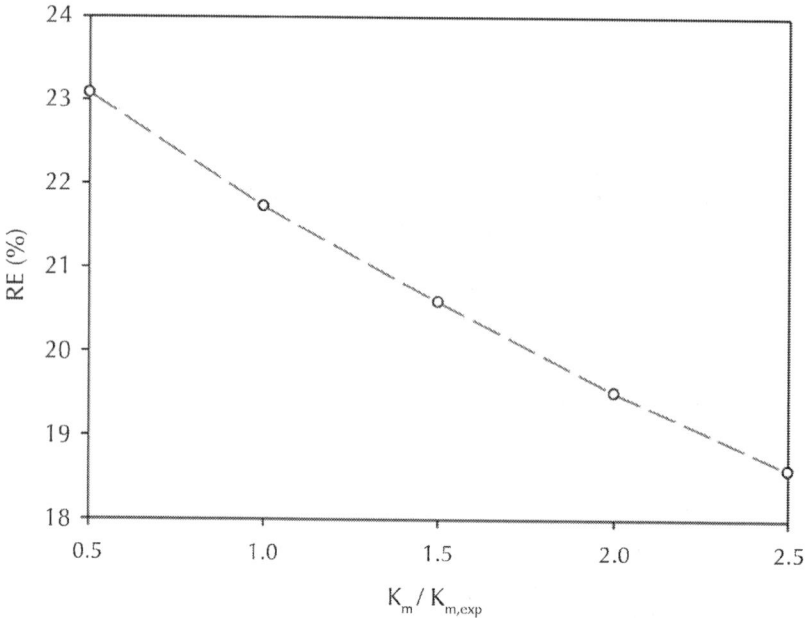

Figure 9: Effect of changing K_m in model 2 (with substrate inhibition) on removal efficiency (RE). Toluene-contaminated air volumetric loading is 35.6 m^3/m^3 biofilter/h, inlet toluene gas-phase concentration is 3.01 g/m^3, $K_{m,experimental}$ is 5.34 g/m^3, experimental data adopted from Park and Jung [12].

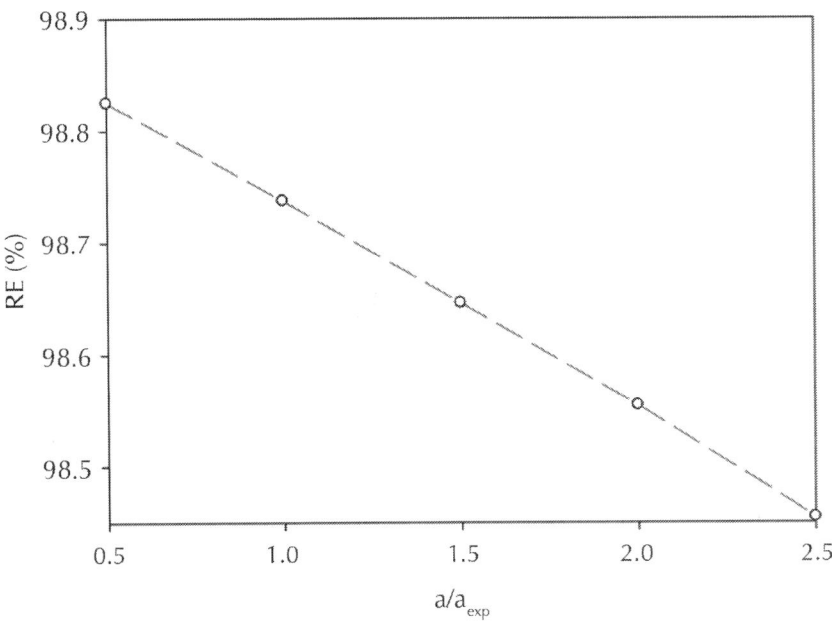

Figure 10: Effect of changing decay rate "a" in model 1 (no substrate inhibition) on removal efficiency (RE). Toluenecontaminated air volumetric loading is 77.6 m³/m³ biofilter/h, inlet toluene gas-phase concentration is 4.6 g/m³, $a_{experimental}$ is 0.0017 h⁻¹, experimental data adopted from Aizpuru et al. [29].

Effect of Inhibition Constant (K₁)

Inhibition is an important factor in model 2 for the prediction of gas-phase toluene concentration along the biofilter height. The effect of changing the inhibition constant (K_I) on removal efficiency in model 2 is shown in Figure 12. An increase in the inhibition constant (K_I), leads to an increase in toluene removal efficiency in model 2. The results from with model 3 were similar. Again, since model 3 is very similar to model 2, the effect of changing K_m will be the same as in case of model 2.

CONCLUSIONS

In this work, important literature on recent biofilter models are reviewed. The toluene biofilter model of Li and De Visscher [22] is solved and compared with a new experimental data of Zamir et al. [28]. The model is in very good agreement with the experimental data. Thus, the model can be used as a good approximation for full scale biofilter design calculations, particularly for toluene removal. A sensitivity analysis of the three models is performed. Maximum specific growth rate, kinetic constant (K_m), and inhibition constants are more sensitive to removal efficiency than decay rate constant. Thus, accurate estimation of these parameters is important.

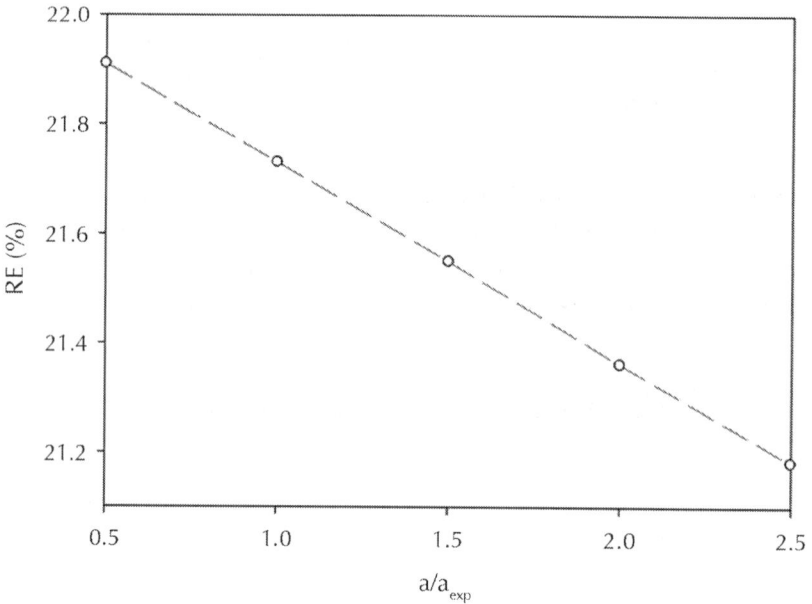

Figure 11: Effect of changing K_m in model 2 (with substrate inhibition) on removal efficiency (RE). Toluene-contaminated air volumetric loading is 35.6 m³/m³ biofilter/h, inlet toluene gas-phase concentration is 3.01 g/m³, $a_{experimental}$ is 0.0017 h⁻¹, experimental data adopted from Park and Jung [12].

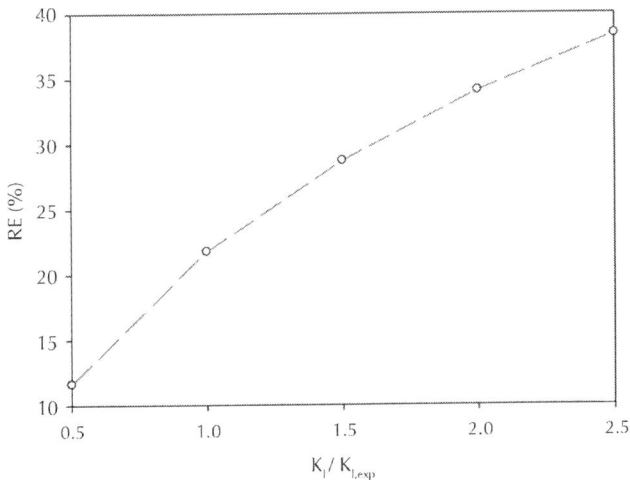

Figure 12: Effect of changing K_I in model 2 (with substrate inhibition) on removal efficiency (RE). Toluene-contaminated air volumetric loading is 35.6 m³/m³ biofilter/h, inlet toluene gas-phase concentration is 3.01 g/m³, $K_{I,experimental}$ is 2.66 g/m³, experimental data adopted from Park and Jung [12].

Although the model does not provide an insight into the nature of the limiting factors such as oxygen effects and diffusion limitations involved in the biofiltration process, the model is simple and will enable designing of biofilters using fewer biofilter parameters.

ACKNOWLEDGMENTS

The authors acknowledge the support of the Chemical Engineering Department at the American University of Sharjah.

REFERENCES

1.   T. Nukunya, J. S. Debinny and T. T. Tsotsis, "Application of a Pore Network Model to a Biofilter Treating Ethanol Vapor," Chemical Engineering Science, Vol. 60, No. 3, 2005, pp. 665-667. doi:10.1016/j.ces.2004.08.038

2.   B. V. Babu and S. Raghuvanshi, "Simulation Studies on Transient Model for Biofilter Operated in Periodic Mode," Journal on Engineering and Technology, Vol. 1, No. 4, 2006, pp. 72-76.

3.   C. R. Soccol, et al., "Biofiltration: An Emerging Technology," Indian Journal of Biotechnology, Vol. 2, No. 2, 2003, pp. 396-410.

4.   Z. Shareefdeen and A. A. Shaikh, "Analysis and Comparison of Biofilter Models," Chemical Engineering Journal, Vol. 65, No. 1, 1998, pp. 55-61.

5.   J. S. Devinny and J. Ramesh, "A Phenomenological Review of Biofilter Models," Chemical Engineering Journal, Vol. 113, No. 2-3, 2005, pp. 187-196.doi:10.1016/j.cej.2005.03.005

6.   M. Hirai, M. Ohtake and M. Shoda, "Removal Kinetics of Hydrogen Sulfide, Methanethiol and Dimethyl Sulfide by Peat Biofilters," Journal of Fermentation Bioengineering, Vol. 70, No. 5, 1990, pp. 334-339. doi:10.1016/0922-338X(90)90145-M

7.   S. P. P. Ottengraf and A. H. C. van den Oever, "Kinetics of Organic Compound Removal from Waste Gases with a Biological Filter," Biotechnology and Bioengineering, Vol. 25, No. 12, 1983, pp. 3089-3102. doi:10.1002/bit.260251222

8.   Z. Shareefdeen, B. C. Baltzis, Y. S. Oh and R. Bartha, "Biofiltration of Methanol Vapor," Biotechnology and Bioengineering, Vol. 41, No. 5, 1993, pp. 512-524.doi:10.1002/bit.260410503

9.   Z. Shareefdeen and B. C. Baltzis, "Biofiltration of Toluene Vapor under Steady-State and Transient Conditions: Theory and Experimental Results," Chemical Engineering Science, Vol. 49, No. 24, 1994, pp. 4347-4360. doi:10.1016/S0009-2509(05)80026-0

10. M. A. Deshusses, G. Hamer and I. J. Dunn, "Behavior of Biofilters for Waste Air Biotreatment. 1. Dynamic Model Development," Environmental Science Technology, Vol. 29, No. 4, 1995, pp. 1048-1058. doi:10.1021/es00004a027

11. M. A. Deshusses, G. Hamer and I. J. Dunn, "Behavior of Biofilters for Waste Air Biotreatment. 2. Experimental Evaluation of a Dynamic Model," Environmental Science Technology, Vol. 29, No. 4, 1995, pp. 1059-1068. doi:10.1021/es00004a028

12. O. Park and I. Jung, "A Model Study Based on Experiments on Toluene Removal under High Load Condition in Biofilters," Biochemical Engineering Journal, Vol. 28, No. 3, 2006, pp. 269-274. doi:10.1016/j.bej.2005.11.011

13. C. P. Yang, H. Chen, W. Qu, Y. Y. Zhong, X. Zhu and M. T. Suidan, "Modeling Biodegradation of Toluene in Rotating Drum Biofilter", Water Science Technology, Vol. 54, No. 9, 2006, pp. 137-144. doi:10.2166/wst.2006.860

14. Q. Liao, X. Tian, R. Chen and X. Zhu, "Mathematical Model for Gas-Liquid Two-Phase Flow and Biodegradation of a Low Concentration Volatile Organic Compound (VOC) in a Trickling Biofilter," International Journal of Heat and Mass Transfer, Vol. 51, No. 7-8, 2007, pp. 1780- 1792. doi:10.1016/j.ijheatmasstransfer.2007.07.007

15. J. Chen, Y. Jiang, H. Sha and W. Zhang, "Dynamics Model for Nitric Oxide Removal by a Rotary Drum Biofilter," Journal of Hazardous Materials, Vol. 168, No. 2-3, 2009, pp. 1047- 1052. doi:10.1016/j.jhazmat.2009.02.159

16. G. Spigno, M. Zilli and C. Nicolella, "Mathematical Modeling and Simulation of Methanol Degradation in Biofilters," Biochemical Engineering Journal, Vol. 19, No. 3, 2004, pp. 267-275. doi:10.1016/j.bej.2004.02.007

17. G. Baquerizo, et al., "A Detailed Model of a Biofilter for Ammonia Removal: Model Parameters Analysis and Model Validation," Chemical Engineering Journal, Vol. 113, No. 2-3, 2005, pp. 205-214. doi:10.1016/j.cej.2005.03.003

18. C. Lu, K. Chang and S. Hsu, "A Model for Treating Isopropyl Alcohol and Acetone Mixtures in a Trickle-Bed Air Biofilter," Process Biochemistry, Vol. 39, No. 12, 2004, pp. 1849-1858. doi:10.1016/j.procbio.2003.09.019

19. K. Chmiel, et al., "Periodic Operation of Biofilters. A Concise Model and Experimental Validation," Chemical Engineering Science, Vol. 60, No. 11, 2005, pp. 2845- 2850. doi:10.1016/j. ces.2004.12.035

20. EPA, "Chemical Summary of Toluene," 2011. www.epa.gov/chemfact/s_toluen.txt

21. K. Singh, R. S. Singh and S. N. Upadhyay, "Biofiltration of Toluene Using Wood Charcoal as the Biofilter Media," Bioresource Technology, Vol. 101, No. 11, 2009, pp. 3947- 3951. doi:10.1016/j.biortech.2010.01.025

22. G. Q. Li and A. De Visscher, "Toluene Removal Biofilter Modeling," Air and Waste Management Association, Vol. 58, No. 7, 2008, pp. 947-956. doi:10.3155/1047-3289.58.7.947

23. A. De Visscher and O. Van Cleemput, "Simulation Model for Gas Diffusion and Methane Oxidation in Landfill Cover Soils," Waste Management, Vol. 23, No. 7, 2003, pp. 581- 591. doi:10.1016/S0956-053X(03)00096-5

24. S. J. Pirt, "Principles of Microbe and Cell Cultivation," Blackwell Scientific, Oxford, 1975.

25. J. S. Devinny, M. A. Deshusses and T. S. Webster, "Biofiltration for Air Pollution Control," CRC Press, Boca Raton, 1999.

26. J. B. S. Haldane, "Enzymes," Longmans, Green, New York, 1930.

27. M. O. Pereira, M. Kuehn, S. Wuertz, T. Neu and L. F. Melo, "Effect of Flow Regime on the Architecture of a Pseudomonas Fluorescens Biofilm," Biotechnology and Bioengineering, Vol. 78, No. 2, 2002, pp. 164-171. doi:10.1002/bit.10189

28. S. M. Zamir, R. Halladj and B. Nasernejad, "Removal of Toluene Vapors using a Fungal Biofilter under Intermittent Loading," Process Safety and Environmental Protection, Vol. 89, No. 1, 2011, pp. 8-14. doi:10.1016/j.psep.2010.10.001

29. A. Aizpuru, B. Dunat, P. Christen, R. Auria, I. Garcia-Pena and S. Revah, "Fungal Biofiltration of Toluene on Ceramic Rings," Journal of Environmental Engineering, Vol. 131,

No. 3, 2005, pp. 396-402. doi:10.1061/(ASCE)0733-9372(2005)131:3(396)

30. L. F. Bautista, A. Aleksenko, M. Hentzer, A. SanterreHenriksen and J. Nielsen, "Antisense Silencing of the CreA Gene in Aspergillus Nidulans," Applied and Environmental Microbiology, Vol. 66, No. 10, 2000, pp. 4579-4581. doi:10.1128/AEM.66.10.4579-4581.2000

31. M. Carlsen, A. B. Spohr, J. Nielsen and J. Villadsen, "Morphology and Physiology of an α-Amylase Producing Strain of Aspergillus Oryzae during Batch Cultivations," Biotechnology and Bioengineering, Vol. 49, No. 3, 1996, pp. 266-276. doi:10.1002/(SICI)1097-0290(19960205)49:3<266::AID-BIT4>3.0.CO;2-I

32. T. D. Reynolds and P. A. Richards, "Unit Operations and Processes in Environmental Engineering," 2nd Edition, PWS Press, Boston, 1996.

33. "EPA Online Tools for Site Assessment Calculations—Estimated Henry's Law Constant-OSWER Method," U.S. Environmental Protection Agency, 2011. http://www.epa.gov/ATHENS/learn2model/part-two/onsite/esthenry.html

34. K. F. Reardon, D. C. Mosteller and J. D. Bull Rogers, "Biodegradation Kinetics of Benzene; Toluene, and Phenol as Single and Mixed Substrates for Pseudomonas Putida F1," Biotechnology & Bioengineering, Vol. 69, No. 4, 2000, pp. 385-400. doi:10.1002/1097-0290(20000820)69:4<385::AID-BIT5>3.0.CO;2-Q

35. R. Mirpuri, W. Jones and J. D. Bryers, "Toluene Degradation Kinetics for Planktonic and Biofilm: Growth Cells of Pseudomonas putida 54G," Biotechnology and Bioengineering, Vol. 53, No. 6, 1997, pp. 535-546. doi:10.1002/(SICI)1097-0290(19970320)53:6<535::AID-BIT1>3.0.CO;2-N

36. Y. B. Choi, J. Y. Lee and H. S. Kim, "A Novel Bioreactor for the Biodegradation of Inhibitory Aromatic Solvents: Experimental Results and Mathematical Analysis," Biotechnology and Bioengineering, Vol. 40, No. 11, 1992, pp. 1403-1411. doi:10.1002/bit.260401115

37. F. X. Prenafeta-Bold, A. Kuhn, D. M. A. M. Luykx, H. Anke, J. W. Van Groenestijn and J. A. M. De Bont, "Isolation and Characterization of Fungi Growing on Volatile Aromatic Hydrocarbons as Their Sole Carbon and Energy Source," Mycological Research, Vol. 105, No. 4, 2001, pp. 477-484. doi:10.1017/S0953756201003719

38. A. Vergara-Fernandez, L. L. Molina, N. A. Pulido and G. Aroca, "Effects of Gas Flow Rate, Inlet Concentration and Temperature on the Biofiltration on Toluene Vapors," Journal of Environmental Management, Vol. 84, No. 2, 2007, pp. 115-122.doi:10.1016/j.jenvman.2006.04.009

39. S. M. Maliyekkal, E. R. Rene, L. Philip and T. Swaminathan, "Performance of BTX Degraders under Substrate Versatility Conditions," Journal of Hazardous Materials, Vol. 109, No. 1-3, 2004, pp. 201-211. doi:10.1016/j.jhazmat.2004.04.001

Citations

CHAPTER 1

Pratesh Jayaswal, A. K. Wadhwani, and K. B. Mulchandani, "Machine Fault Signature Analysis," International Journal of Rotating Machinery, vol. 2008, Article ID 583982, 10 pages, 2008. doi:10.1155/2008/583982.

CHAPTER 2

Cesar R.F. Azevedo, J.C.K. das Neves, A. Sinatora, Failure analysis of belt/roll tribological pair used for the production of eucalypt fiber panels, Engineering Failure Analysis, Volume 15, Issues 1–2, January–March 2008, Pages 165-181, ISSN 1350-6307, http://dx.doi.org/10.1016/j.engfailanal.2006.11.016.

CHAPTER 3

Riyas Sharafudeen, The Manufacturing Process Parameters Affecting Color and Brightness of TiO2 Pigment, doi: 10.1186/2228-5547-3-26.

CHAPTER 4

GuoRong Wang, LinYan Chen, Min Zhao, Rong Li, BenSheng Huang, The research on failure analysis of fluid cylinder and fatigue life prediction, Engineering Failure Analysis, Volume 40, May 2014, Pages 48-57, ISSN 1350-6307, http://dx.doi.org/10.1016/j.engfailanal.2014.01.007.

CHAPTER 5

G Das, A.n Sinha, S.K.Mishra (Pathak), D.K Bhattacharya, Failure analysis of counter shafts of a centrifugal pump, Engineering Failure Analysis, Volume 6, Issue 4, 1 August 1999, Pages 267-276, ISSN 1350-6307, http://dx.doi.org/10.1016/S1350-6307(98)00037-5.

CHAPTER 6

Tae-Gu Kim, Hong-Chul Lee, Failure analysis of MVR (machinery vapor recompressor) impeller blade, Engineering Failure Analysis, Volume 10, Issue 3, June 2003, Pages 307-315, ISSN 1350-6307, http://dx.doi.org/10.1016/S1350-6307(02)00077-8.

CHAPTER 7

Amir Mostafaei, Seyed Majid Peighambari, Farzad Nasirpouri, Failure analysis of monel packing in atmospheric distillation tower under the service in the presence of corrosive gases, Engineering Failure Analysis, Volume 28, March 2013, Pages 241-251, ISSN 1350-6307, http://dx.doi.org/10.1016/j.engfailanal.2012.10.028.

CHAPTER 8

Sr an M. Bošnjak, Dejan B. Momčilović, Zoran D. Petkovi , Milorad P. Panteli , Nebojša B. Gnjatovi , Failure investigation of the bucket wheel excavator crawler chain link, Engineering Failure Analysis, Volume 35, 15 December 2013, Pages 462-469, ISSN 1350-6307, http://dx.doi.org/10.1016/j.engfailanal.2013.04.025.

CHAPTER 9

Laura R. Jarboe, Xueli Zhang, Xuan Wang, Jonathan C. Moore, K. T. Shanmugam, and Lonnie O. Ingram, "Metabolic Engineering for Production of Biorenewable Fuels and Chemicals: Contributions of Synthetic Biology," Journal of Biomedicine and Biotechnology, vol. 2010, Article ID 761042, 18 pages, 2010. doi:10.1155/2010/761042.

CHAPTER 10

Muhammad Qasim and Zarook Shareefdeen, Analysis of a Recent Biofilter Model for Toluene Biodegradation, http://dx.doi.org/10.4236/aces.2013.31006.

Index